PRAISE FOR *The Women of the House*

"This extraordinary story of an American dynasty founded
and perpetuated by women will be a valuable addition
to both colonial and women's history collections."
—*Booklist*

"Zimmerman's prodigious research unearths
a mother lode of data on colonial American women."
—*Publishers Weekly*

"Lively and informative . . . Jean Zimmerman's extraordinary
research and energetic writing help readers better understand and
appreciate the roles played by women during [the colonial] era."
—*BookPage*

"A dense, richly textured portrayal of New York's founding,
that neither sugarcoats nor flinches from the
unsavory aspects of much of its historical inheritance."
—*Barnard* magazine

"Zimmerman gestures toward larger themes in social history.
She notes that married women's property rights were eroded
when the English took over New Amsterdam, and she
touches on the dangers that Loyalists in New York
faced during the Revolutionary War."
—*Kirkus Reviews*

PRAISE FOR
Tailspin: Women at War in the Wake of Tailhook

"An intelligent, inside look at one of the most
embarrassing scandals in naval history . . . An excellent read."
—*The Village Voice*

"Zimmerman's Tom Wolfe–esque 'new journalism' style,
as heavy on sights, sound effects, jargon and description as it is
on 'facts,' really works here to flesh out life as it is,
not as the official documents say it is."
—*The Baltimore Sun*

"*Tailspin* is a stingingly brisk read, every bit as crisp and salty
as the top-gun culture it dissects. Zimmerman deftly skewers
the frat-boy mentality that led to Tailhook and—on the bright
side—introduces the military women who are our real heroes."
—Barbara Ehrenreich

PRAISE FOR *Raising Our Athletic Daughters*

"Captures the spirit of all-star athletes and today's sports-loving
girls who want to grow up to be strong, capable women.
It's the jock version of *Reviving Ophelia*."
—*USA Today*

The Women of the House

ALSO BY JEAN ZIMMERMAN

Made from Scratch: Reclaiming the Pleasures of the American Hearth

Raising Our Athletic Daughters:
How Sports Can Build Self-Esteem and Save Girls' Lives
(with Gil Reavill)

Manhattan (with Gil Reavill)

Tailspin: Women at War in the Wake of Tailhook

The Women of the House

HOW A COLONIAL SHE-MERCHANT BUILT A MANSION, A FORTUNE, AND A DYNASTY

Jean Zimmerman

A Harvest Book
Harcourt, Inc.
Orlando   Austin   New York   San Diego   London

For Gil and Maud

www.HarcourtBooks.com

The Library of Congress has cataloged the hardcover edition as follows:
Zimmerman, Jean.
The women of the house: how a colonial she-merchant built a
mansion, a fortune, and a dynasty/Jean Zimmerman.—1st ed.
p.   cm.
Includes bibliographical references (p.) and index.
1. Philipse, Margaret Hardenbroeck.   2. Philipse, Margaret
Hardenbroeck—Family.   3. Phillips family   4. Women
merchants—New Netherland—Biography.   5. Merchants—
New Netherland—Biography.   6. New Netherland—
Biography.   7. New Netherland—Commerce—History.
8. New Netherland—Social conditions.   9. New York
(State)—History—Colonial period, ca. 1600–1775—Biography.
I. Title.
F122.1.P48Z56   2006
974.7'1020922—dc22   2005027992
ISBN 978-0-15-101065-3
ISBN 978-0-15-603224-7 (pbk.)

Text set in Bembo
Designed by Lydia D'moch

Printed in the United States of America
First Harvest edition 2007
A C E G I K J H F D B

Contents

Prologue: 1685

The beginning of April in the Hudson River Valley is marked by a hard bluish light in late afternoon, haunting as it is beautiful. By day the sunshine makes mud from dirt patches amid the melting snow, while each night the mud cools again, freezing to crunchy toast beneath the hooves of the spotted deer as they travel through the moonlight like big-eyed ghosts. The tangled thornbushes that divide wood and field color up suddenly, a burnished amethyst red. Along the shoreline, grasses taller than a man hold on to their pale, dead yellow as if a warmer season will never arrive.

Spring will inevitably come. But in this interim, anticipation rises as surely as the sap in the trees that stretch up their cold branches to the early April sun. Anticipation throbs in the air of the Hudson Valley like a pulse: Soon, the fur hunters will emerge from deep in the frozen woods. And the time of trade will begin.

Seated on her wooden stoop, wrapped in an ankle-length cloak, a silver mink muff on her lap, Margaret Hardenbroeck lifts a long-stemmed clay pipe to her mouth and surveys her landscape in the chilly dusk. To the west she can see the wide river fringed by the yellow reeds of the salt marsh. Much of the marsh lies in the deep shadow of soaring rock columns on the opposite shore, a majestic geological feature that will be known to future generations as the Palisades. The brim of Margaret's peaked castor hat shields her eyes from the late-day sun's glare over that ridge. Though the surface of the river has nearly melted, gray cakes of ice still nudge the shoreline, and the center of the water ripples sluggishly with its opaque hazel glow.

Leaning against the doorframe, she drinks her pipe in the Dutch manner, gulping the hot smoke. Her West Indian horse snorts quietly beside her, tied up to a tree whose crown in the summer months throws a cooling shade over the whole house. Margaret's back aches from the day of riding—now almost fifty, she has curls gone to gray that stray in a loose frizz around her face, whose lines have coarsened from years of baking sun and sea spray. Her house stands alone, the only human habitation amid the towering pines. Though secluded, it is hardly a quiet spot. The waters of a river the local natives call Nepperhan rush in front of the house, crashing over five waterfalls before flowing into the depths of the deeper, more sober Hudson. Then there is the abrupt shrieking warble of a hawk over the marsh, and she watches it drop like a rock from the sky, so close she can see its tensed feet with their ink-black talons.

Earlier that day, Margaret had taken her horse along the Broad Way, the ancient path that followed an irregular course across the body of Manhattan, plunging through its deepest woods and between the soaring glacial boulders at the top of what would one day be Central Park, finally climbing the northern ridge where the remaining Weckquaesgecks of Manhattan had retreated recently after too many years of violence at the hands of colonists. She'd seen skunk cabbage sprouting in the shady creek beds and a few tufts of onion grass where the full sun hit. Since the trees had yet to leaf she could peer down through the weave of gray branches to the wide rivers on either side of the island. The leaves of the wild strawberry, exposed now that the snow had melted, blanketed the forest floor. The fruit was not an exotic treat for Margaret and her family, but along with fiddlehead ferns and wild ramps found a central place on their table in springtime, when fresh produce was rare. By April the larder in nearly every household held mainly dried apples or withered turnips.

Settlers had launched a successful hunt to kill the wolves that terrorized this lonely wilderness at the north end of Manhattan. Still, she was glad she brought her musket. In a broad clearing punctuated by rotting knee-high stumps she passed the ruins of Dutch homes, their crude timbers abandoned dozens of years before when an early community of settlers fled back down-island to escape the threat of massacre. It had now been revived nearby as the bustling settlement of Haarlem.

Just beyond, she reached a landmark known as the Wading Place, a short stretch of the creek that connected the Harlem River to the east of Manhattan with the Hudson to the west. The creek was all that separated the island from the mainland to its north. Its Dutch-derived name, Spuyten Duyvil—Spit at the Devil—suggested the lethal properties of the stream, which would one day be diverted, widened, straightened, and deepened to become the Harlem River Ship Canal. For thousands of years the native people forded this water en route from their rock shelters at the top of the island to the northern lands where they hunted and raised crops, and for Europeans and Indians alike in Margaret's time it is still the one route north from or south into Manhattan. There is no other path. No one has ever built a bridge over the creek, and one entrepreneur's attempt to launch a ferry service had been stymied by colonists' resistance to the idea of paying for a crossing that should come free. Passage now, as ever, can be made only at low tide, when the clear water does not rise above three feet; then, looking down to the bottom, the smooth river stones sparkle like cut gems, interspersed with succulent oysters for the taking.

Sidesaddle, Margaret splashed across. It was not even approaching high tide with its deadly currents, but the water soaked her shoes and stockings and the hem of her skirt. The coarse linen riding petticoat that guarded her dark satin gown was already splattered with mud from her trip to the top of Manhattan. But that didn't matter. Soon after her horse heaved up the far bank she would reach her own land, and in a matter of hours she would arrive at the house she had built.

Now, as the sun drops behind the blackening cliffs, Margaret stuffs her hands in the downy muff. She has laid a fire on the hearth with an oak log so fat it would heat the thick-walled *kamer* for days. Here at her house she would walk, and sleep, and have her meat, her rasped cheese and bread and the jar of beer. The hawks over the salt marsh would soar, scream, fall on their prey. With the river always in sight, the fur trader would wait. For now, the raw air and cold soil hold tight to their promise of riches.

*B*y the close of her life, Margaret Hardenbroeck would be the richest woman in what had become the English province of New York. She was unquestionably the first to have built her own fortune.

In the province's remote northern woodlands she bartered with native hunters for consignments of the finest fresh-killed animal hides. She amassed a fleet of trade vessels to convey the pelts from the New World to Europe, where a mania for fur ensured her ever-expanding fortune. To assure quality control, Margaret took to the seas on her own ships, traveling from North America to Europe as supercargo—the person charged with the authority for all merchandise stowed on board. By the end of her life, her assets included multiple properties in Manhattan, land in New Jersey, a mansion in Albany, and an estate on the money-minting island of Barbados, from whose ports her ships hauled sugar and returned with beef, horses, and enslaved Africans.

The house on the Nepperhan, though, was the heart of Margaret's empire. This deceptively modest outpost was eventually surrounded by tens of thousands of acres of prime farmland she acquired north of Manhattan, an estate that comprised much of what is now Westchester County, New York. The fortune she built, passed along down the years like a solid-gold relay baton, would give the successive generations of women in her family the resources to lead remarkable lives of their own.

Margaret's successor would be Catherine, a New York–bred heiress who would marry Margaret's widowed husband and direct her own considerable energies to the good work of church building. Next came Joanna, a horsey society doyenne who put her wits to use as the ultimate political wife during an era of slave and civic unrest. Finally there would be Mary, the It girl of mid-1700s New York, who threw over a young war hero named George Washington for a more manageable milquetoast Briton and tried to finesse the tumultuous Revolutionary era by alternately inviting Americans and Tories to tea.

Each of these distinctive and independent women would make their own imprint on the Hudson Valley house that Margaret built. Their renovations, remodelings, and additions gradually transformed the place from an unprepossessing storehouse to the stately mansion known as Philipse Manor Hall. Each generation applied its mark to the project, expressing their most vital ideas and their deeply held values in the dimensions of the house, and in turn their choices reflected the wider and constantly changing culture of this extraordinary period of American history.

We don't have to imagine the culmination of this century-long reign of designing women; it was captured in a sepia drawing that was discovered in 1905 in a Philadelphia print shop. Experts believe the picture, the earliest discovered of the Manor Hall, was used as a real estate circular to convey the attractions of the property when it was parceled off in the wake of the American Revolution.

The drawing shows the house from the perspective of a cliff directly south of the structure, facing it from across the Nepperhan. We see a waterfall splash over a milldam to midstream rapids that seethe with foam. On the far side of the river stand a trio of gristmills; their overshot wheels process the hundreds of pounds of grain that pass through their hoppers every day. From the mills the river hurtles over more falls, then cuts down to the Hudson, where a sloop plies the current, gliding so close to the orchard on the bank below the house that it looks as if a sailor could stick out his hand and pluck a Pippin or a Sweet, two favorite apples of the day.

On the far shore we see the south face of the mansion, its windows placed to receive the warming light of the late afternoon sun. The northern wing of the house is hidden behind a mass of trees. The only people in the frame, occupying a central place in the foreground, are four stylized reclining figures, voluptuous and smiling. They seem embodiments of a conviction that this place has the look and taste and smell of paradise.

$\mathcal{W}$omen such as Margaret Hardenbroeck are conspicuously absent in most accounts of the colonial era. In *A Room of One's Own,* Virginia Woolf deplored the little we know about how our female forebears spent their days. Wives and mothers have always carried the load of feeding families and minding children. But at the same time, women of generations past often had lives more complex, more active, and less defined by their gender than is generally assumed.

We hold an image of the colonial era as the primary domain of men who were trackers and trappers, soldiers and farmers, pioneers and politicians, merchants and sailors and roll-up-your-shirtsleeves laborers. The role of women in this tapestry received scant attention until relatively recently, when scholars of the 1970s spurred much worthy research into the female inhabitants of the English colonies. Even so, historians tend to focus on the handful who have become iconic figures: Anne Hutchinson, Anne Bradstreet, Phillis Wheatley, Martha Washington, Abigail Adams. A worthy crew, but incomplete.

Margaret Hardenbroeck and the women of her dynasty deserve a place at that table; each exemplified the forceful character and individuality that would become the hallmark of American women.

Clearly, Margaret's bravado proved key to her enormous success. But she was lucky, too. She had inherited an egalitarian Dutch tradition, which made such success not only legally possible but also socially acceptable, and had landed in a new country in the process of inventing itself, a place where Old World rules did not always apply. No matter how the climate changed for New York women with British rule, Margaret had secured her fortune and her empire, and she would endow her heirs with the power to do what they chose with their lives.

What follows will, I hope, give an idea of how four very well-off, very idiosyncratic women interacted with a world that has long since disappeared. What it was like to inhale the pungent scent of the first-growth pines, to stroke a thick, matted beaver fur, to taste the fizz of home-brewed cider. What it might have been like to prepare a turtle for roasting or to consume a slick, shucked oyster the size of a dinner plate. How these pleasures coexisted with other, more difficult realities, like the disintegrating bodies of conspirators swaying from the gallows after New York's devastating slave revolt of 1741.

And, finally, to convey the sweat on the brow of a hardworking merchant after she'd finagled a deal in the broiling summer heat of the maritime city.

Part One

1659–1691
MARGARET

I

Her New World

$\mathcal{A}$t the bow of the yacht, her fingers gripping the rail, a young woman stood face into a wind that had buffeted the ship for days with a cocktail of fresh-mown hay, pine sap, and even the sweetness of wildflowers. Behind her, the pilot leaned all his weight on the tiller to thread the vessel through the Narrows, the Hoofden, where even the most seasoned skippers had been known to founder their ships on knife-sharp shoals. The little ship skimmed across the bright open lake called the Upper Bay, and then made its way through one final channel. Finally, a harbor town materialized all at once out of the haze, still a musket shot away but close enough to make out the fort, towering above everything, and the sparse forest of masts in the roadstead before it, colored pennants drooping in the still summer air. As the ship pushed closer, the dense heat of the land descended upon the deck like a wet sponge.

The woman at the rail spotted a windmill's sails turning counterclockwise above a church spire. Over the cry of gulls she heard the bellow of sea lions that sprawled across the black rocks at the island's arrow-shaped foot. The rooflines of the port rose into view, then behind them a scattering of farms tucked into rolling hills, their fields interspersed with stretches of forest. A white signal flag flew above the fort to show that a ship had safely reached the harbor. As the guns began to crack out their welcoming salutes, children could be seen running barefoot toward the shore.

The captain shouted to make the ship fast. A half-dozen small boats, a sturdy raft, and a canoe pushed away from the town's long Winebridge Pier, built just this year. Carrying port inspectors and merchants, the boats headed toward the ship to ferry passengers and goods ashore. A frill of seawater washed the pebble beach north of the dock. Farther north still lay raw pastures, sand hills, and salt marshes, acres of them, all along the island's coast. An immense flock of bluebills splashed down nearby. Some River Indians stood beside their canoes, waist-deep in the water, unloading nets of oysters they would hawk door-to-door to housewives, who would pickle the meat in jars for export to the planters of the West Indies.

The year was 1659. The woman at the ship rail was Margaret Hardenbroeck, she was all of twenty-two years old, and this was her New World. The seaport before her was tiny compared with the European ports of Amsterdam or London, but it was a promising entrepôt, an infant marketplace that just might grow to be a money-making giant. Holland had the audacity to christen this settlement, which thirty-five years ago was nothing but a bare-dirt trading post, after its urbane commercial capital, Amsterdam. Now, finally, the frontier community of New Amsterdam was beginning to look as if it might amount to something.

New Amsterdam was not only a market center. It also was the consummate company town. The company was the Dutch West India Company, an entity controlled by a collection of prosperous burghers who persuaded the Dutch government to grant them a monopoly on trade with West Africa and the Americas and the right to colonize territories. One such territory included the pristine slice of land that ran south from present-day Albany through the island of Manhattan. New Netherland encompassed lands on either bank of the Hudson, as well as choice sections of what later would become New Jersey, Delaware, and Connecticut. Dutch colonists were scattered through Manhattan, Long Island, Brooklyn, and Staten Island, as well as up the Hudson River, where the town of Beverwyck (later known as Albany) was a nucleus for the growing communities of Schenectady, Catskill, and Wiltwyck.

Whether colonists arrived as employees of the Company, sold it the products of their land, or shopped for tools at its store, they all depend upon it for survival. In exchange, though, the Company had always taken care to provision its colonists—unlike, say, the English, whose ill-equipped settlers first landed in Virginia in 1607 and, faced with famine, choked down snakes, leather boots, and sometimes each other. To the Dutch, food mattered. In 1625, immediately after the first vessels reached Manhattan, three ships followed with more than one hundred head of hogs, sheep, cows, and horses destined for Company farms. En route, each animal had a private, sand-cushioned stall and an individual handler who "attends to it and knows what he is to get if he delivers it alive." There would be no "starving times" for New Netherland.

Close by the waterfront, on Winkelstraet, or Shop Street, the Company built a full block of brick warehouses, five under one roof, which it supplemented with a cavernous packinghouse that commanded a perfect view of all harbor traffic. Other nearby warehouses belonged to the town's most successful private merchants. These buildings functioned in the same way as those that crowded the ancient seaports of Holland. With their stately red-roofed facades, they would easily have fit in on the Heerengracht, the grandest canal in Amsterdam. Their imposing heft appeared somewhat discordant in a town that had just finished cutting stones for its first paved street. But that did not matter. The colony's commercial drive would not be thwarted by a lack of refined conditions.

On Manhattan, at the very heart of the Dutch colony, every single thing—from cloth to seashells to human flesh—possessed a dynamic cash value, forever changing, always in play. And for that reason the little town of New Amsterdam at the island's southern tip vibrated with a fast and loose, somewhat chaotic energy that was absent at more established ports. Small-scale shipyards and workshops thrived along the water's edge. In a square beside the fort, a new city market drew shoppers every Saturday to buy the ripe produce the "country people" brought by cart and ferry from their Brooklyn and Harlem farms, or *boeweries*. New Amsterdam was a polyglot town: Eighteen languages had been counted here by the time Margaret arrived in 1659. Business was good.

*D*espite the economic vigor of New Amsterdam, development of the colony as a whole had been erratic—to the irritation of the West India Company's directors. From the start, the Dutch had been beleaguered in North America; the French to the north and the British to the east sought not only to appropriate its resources but also to oust its colonists and capture the entire franchise. The English posed a particular threat. They galled the Dutch by installing well-fortified villages in the loamy valley of the Connecticut River, uncomfortably close to, and even within, the bounds of territory the Dutch considered irrefutably theirs.

Margaret was standing at the bow of her ship thanks in part to this irritation. The West India Company had renewed its commitment to settling the colony with permanent inhabitants, not just cut-and-run traders, reasoning that enemies would have a tougher time grabbing tenanted farms than land that lay fallow. The Company circulated leaflets such as the eighty-four-page "Short Account of New Netherland's Potential Virtues," a heavy-handed pitch for the new American Eden. If not seduced by milk and honey, potential settlers might be intrigued by offers of farm acreage, house lots, and military protection. Any man who could prove his ability to earn would get a free ride over, his family, too. Finally, in the 1650s, streams of Hollanders, including families with children, had begun to make the transatlantic leap of faith. The bare-dirt settlement was becoming a real Dutch community.

Margaret represented another increasing demographic of New Netherland: young, single women who crossed the ocean for opportunity in America. It sounds about three hundred years too soon, but Manhattan had begun to welcome autonomous, self-sufficient women to its shores. Dutch ship rosters from the 1650s cite a sharp rise in the number of "maidens" (the locution for any single woman, whether she was never married, divorced, or widowed) berthed alongside more traditional immigrant types, the men who had crossed the Atlantic alone or with their families since the founding of the colony.

How dramatic was the sudden influx of independent women? The exact count of female pioneers will never be known, because passenger records are fragmentary. Ship rosters of the time often omitted immigrants with prepaid fares, a system that relegated many men and women to historical invisibility. However, available records show an unmistakable trend. On fifteen documented trips from Holland to New Netherland between 1630 and 1644, only 2 of 104 identifiably single passengers were female, and they were adolescent sisters bound for a reunion with their father. A generation later, the proportion of single women named on ship lists had climbed to fully 6 percent of arriving immigrants, a statistical event that had no counterpart among all the other colonial expansions of the seventeenth century.

To appreciate the increasing presence of unattached women on the streets of the Dutch settlement, consider a sampling of rosters for Amsterdam–Manhattan voyages in the three years before

Margaret's arrival. Out of ten ships, seven carried at least one of these free agents. In May 1658, for example, a maiden named Tryntje Pieters made her passage on the *Gilded Beaver.* Four weeks later, Jannetje Hermens departed Amsterdam on the *Brownfish.* Nine single women crossed on the *Faith* in February 1659. That same year, the *Beaver* carried three passengers identified as maidens: Annetje Ruytenbeck, Geertry van Meulen, and Maintien Jans.

And with them, Margaret Hardenbroeck? Margaret was one of the many immigrants whose names do not appear on any manifest. When this particular maiden pushed off from the pier in Holland, she presumably had a paid-in-full ticket to ride. If Margaret made her crossing on the *Beaver*—a likelihood, given the year and season of the voyage—her shipmates would have included a wheelwright with his wife, baby, and servant; six French farmers; a tailor; and a hatter. Margaret would have joined Annetje, Geertry, and Maintien in the unprecedented historical wave of maidenhood to reach the New World.

And yet something set Margaret apart from her fellow shipmaidens. She arrived in America with a job already in hand—a position to rival that of any wheelwright or hatter aboard her ship. Margaret was a "factor," a representative for a trading business. Her cousin, a well-to-do merchant named Wouter Valck, owned the business, and it would have fallen to him to pay his employee's passage to Manhattan—the probable reason her name will never be found on a roster.

Margaret was born into a trading family—not petty merchants, dispensing dry goods such as scrubbing brushes or thread out of a cramped shop front, but ship merchants. Wouter Valck was her family's success story. He booked space on transcontinental vessels to move the bulk goods he bought and sold. And although Margaret's father Adolph, brothers Johannes and Andries, and uncle Pieter all worked in overseas trade, buying and selling in some quantity, the less-experienced Margaret was the one Valck had anointed to deal on his behalf. As his factor, she was assigned to collect money from his customers and seek out buyers for merchandise he shipped from Europe to America. In Holland she spent the bulk of her workday paying calls on the merchant's accounts. Up close and personal was the preferred way to conduct this business: showing up unannounced at a debtor's shop or residence, becoming a nuisance on his doorstep to collect the monies owed. The role of the heavy was an integral part of seventeenth-century commerce, Dutch-style.

Margaret Hardenbroeck, Valck had discovered, was born to it: More than brash, she was a brute in silks. During six years of employment with Valck, both before and after her 1659 voyage, Margaret proved her worth in transaction after transaction. She never shrank from representing Valck in court, either as plaintiff or defendant. Equally important, Margaret was studious in keeping her boss informed about his business. In her correspondence, Margaret executed a legible hand in an era when the quality of a person's cursive trumpeted character and breeding. She also possessed a forceful bearing and public-speaking skills—attributes that were hardly the norm for women of seventeenth-century Europe, where business had a generally masculine flavor.

In Holland, however, to the great material advantage of Wouter Valck, the norm was the reverse.

Another term that characterized Margaret at twenty-two, besides *maiden*, besides *factor*, was one that working women of her century employed to describe themselves. With every fiber of her canny, thrusting, independent being, Margaret Hardenbroeck was a *she-merchant*. The term appeared elsewhere in Europe and followed settlers to the New World, but in Holland the concept was ingrained as it was nowhere else. The people of the Netherlands accepted women as she-merchants, respecting their skills and their strengths in commerce, providing female Hollanders like Margaret with a sturdy platform from which they could aggressively court success in trade.

She-merchants could be found among Dutch tavern owners and fish sellers, laundresses and lace makers. In villages she-merchant hucksters cruised the streets to hawk strawberries or roasted nuts. Factors generally were male. And yet it was not considered preposterous for Margaret's cousin to entrust a young, relatively inexperienced woman with advancing his business. Margaret's extraordinary aptitude was undergirded by a progressive legal system that gave Dutch women marital and inheritance rights unheard of in other countries, social mores that sanctioned greater independence outside the home, and universal elementary-level education for girls as well as boys. An anonymous seventeenth-century English writer, publisher of a tract that advocated greater respect for the female sex, assessed the shrewd dealings of women in Holland. Unlike their counterparts in England, they could be seen "making and receiving all Payments as well great and small, keeping the Books, balancing the Accounts, and doing all the Business."

Holland exported that entrepreneurial activity to its North American colony. Census figures show that 134 female traders resided in New Amsterdam from 1653 to 1663. Farther north, in the settlement that would come to be called Albany, a recorded forty-seven women worked as traders out of a population of between one and two thousand. New Amsterdam had fifty women employed as brewers, launderers, and bakers during the same period. And upriver, more than a dozen women ran some form of commercial business. Clearly, the female population of New Netherland, just like the women of Patria, the fatherland, did not concern themselves exclusively with tending kitchen gardens, sweeping front steps, and minding children.

Approaching the shore of her new world, Margaret might have looked like an ordinary young woman of her time, with the fair skin Dutch women were famous for, lace shrouding her throat, and sausage curls dangling in front of her ears, but she would not be defined by her appearance any more than she would be circumscribed by her role as wife or mother. The dynamic capitalism that would come to define New York in the centuries to come had deep roots in a 1600s cohort of strivers from Holland whose ranks prominently featured women. Their names might have faded into obscurity, but she-merchants helped give the town its mercantile muscle. One could comfortably operate in the twenty-first century canyons of Wall Street, armed with a briefcase and a cell phone rather than a bunch of iron keys to unlock her warehouse and money box.

$\mathcal{N}$ow the merchants approached Margaret's ship in their small boats and scaled the ladder to the quarterdeck. Like most well-to-do male Europeans, they wore clothes that distinguished them as the swashbuckling sex symbols of the age, equal parts simpering and rogue, with petticoat breeches dribbling ribbons at the knee, slashed sleeves, high-heeled boots, and burly brocaded doublets. Immediately they began to pry open the wooden casks that held bolts of Holland-produced cloth the Indians would accept in trade for skins or even land. The merchants scrutinized the hooks and eyes they commissioned—on credit, with payment dependent upon the quality of the delivery—and plunged their beefy hands into the hogsheads of seed. They counted the briny oaken pipes of Madeira, imagining the wine's flavor, born of lava humus in the waters off Portugal and ripened in the constant heat and rocking of the trip across the Atlantic. They fingered the parcels of blown lace. The bawling Frisian heifer in her tiny stall suffered her eyes and teeth being probed before the farmer come to collect her helped winch her off the deck into his jolly boat; once ashore, his *milch* cow would accompany him to his *boewerie* on foot. Finally, satisfied that their goods were first quality, the merchants descended the companionway to drink brandy in the skipper's cabin.

By this time Margaret and the other passengers had grown impatient to disembark. Behind her in the waist of the yacht paced her shipmates, the farmers, the hatter, the maidens, the wheelwright whose wife cradled their tiny suckling child throughout the crossing. Some of her shipmates were refugees—from hunger, from religious persecution—but most crossed the sea in quest of adventure or a better life for themselves and their families. Everyone was anxious to escape life at sea. For nearly two months the ship had ploughed its way from Amsterdam, to the Canary Islands, to the Indian Islands, to Falmouth, to Virginia, and finally to New Amsterdam, accumulating more merchandise at each stop. Stuffed into low-ceilinged quarters, the passengers—many of whom had slumbered on feather beds at home—tossed in filthy coarse-woven hammocks. Margaret's skin wore a crust of grime. Her hair lay thick with sea salt, stiff as glue beneath the linen cap that was a standard element of a Dutch maiden's costume. The fresh food had long since spoiled. Passengers had grown accustomed to gray beans, weak beer, and biscuit alive with weevils. Many were sick and had spent days crouching over the one head in the deck of the hold, a rough hole beneath which the ocean sucked and swirled.

They were bored, too, from weeks of having little to do. Marvels visible in the skies and sea—star showers, moonbows, tumbling schools of dolphins, the wake of the boat glowing from below as if lit with candles—offered only brief distractions during the voyage. These visions were already forgotten as the colonists pressed forward to set their feet on land. Now they would see for themselves the wonders extolled in pamphlets, letters, and tales of returning sailors. Finally, they would begin to discover what fate had in store for them. Lofted across the world by a set of vague ambitions, many of Margaret's shipmates would find their fortunes trimmed like the wick of a candle, whether by a dearth of talent, a lack of bravado, or a simple willingness to settle for the deal that's offered.

Margaret Hardenbroeck, however—planted solidly at the bow, fixing her gaze on the landscape before her—did not believe in settling. Entering the port of Manhattan, the she-merchant scanned the spectrum of opportunities that stretched from one end of the horizon to the other. She considered the potential in the well-timbered hills rising above the town, the lush valleys and clear, fast creeks, the level maize lands, and the vast woods filled with fat deer for the taking. She imagined what easy alchemy it would take to squeeze the honey out of this land, to transform that honey into gold.

The only question was where to begin. Hitching up her petti-coats, Margaret stepped ashore at Manhattan Island in the Dutch colony of New Netherland, where she would create her own destiny.

2

A Map of Manhattan

*I*t would be understandable for a person who had just survived a two-month ocean passage to occupy herself with practical concerns. Finding a room, say, or washing away the grime of the voyage. But as Margaret arrived at the doorstep of the New World, she already has an assignment from her cousin-employer, Wouter Valck, and Margaret was not the type to waste time getting to work.

Valck wanted her to follow up on a transaction gone bad. He had sold two hogsheads of vinegar to an accountant named Hendrick Jansz. van der Vin in exchange for thirty-two beaver pelts. Valck delivered the goods, collected a down payment of eight pelts, and returned to Amsterdam. His customer had failed to pay the rest as promised. So Valck left an IOU for the remaining twenty-four pelts "in the hands of Margriet Hardenbroeck," soon to sail for America, and he also gave his young factor forty-two liters of olive oil and a swatch of pins to sell on his behalf, according to a statement filed in the Notarial Archives of the Municipal Record Office in Amsterdam. The document, a testament to the Dutch passion for paperwork, is the she-merchant's first recorded presence in the New World.

In five years' time Margaret would have secured her reputation as an independent trader—a "free agent of New Amsterdam," according to the locution of court records filed in 1664 that characterize her standing before a judge. From her first days into the foreseeable future, she would put her all into her work in the colony of New Netherland. And Margaret's all would reliably be a bit more energetic than any of her peers.

Margaret was hot and weary after her long months on the ocean. But she would never be a shirker.

So the she-merchant Margaret set off along Pearl Street.

*L*anding at Manhattan had hardly changed since 1624, when the first settlers reached New York. They were Walloons, French-speaking Belgian Protestants seeking a haven from religious persecution. The English had rebuffed the Walloons' plea for sanctuary in the Virginia colony of Jamestown. Dutch authorities, though, priding themselves on their tolerance for all denominations, welcomed the refugees. Of course, the Dutch also needed warm bodies to settle their new land. After simply swearing an oath of allegiance to the West India Company and the government of the United Provinces, these first New Netherland immigrants settled in. Their welcome set a precedent for the city that became known as a melting pot (*huts pot,* in Dutch) of the world's refugees and seekers.

Eighteen of these families made their way 150 miles north up the Hudson River, where Dutch traders already had constructed a trading post and a military "castle" on a sodden river island that slid into the drink with every spring flood. The rest of the settlers took off to install trading posts along the Connecticut and Delaware rivers. Just eight men remained to assemble their bark huts on Nut

Island (later, Governors Island), a favorite Indian fishing camp that lay a half mile off the tip of Manhattan. The West India Company saw this quarter-mile-square Shangri-la as a good locale not only for harvesting the nut meats that inspired the island's name but also for transferring valuable trade goods between the Company's multiton oceangoing vessels and its smaller, shallow-draft coastal sloops.

Manhattan held only the remains of a small garrison built when independent Dutch skippers began their first trading treks to the region in 1615. Once colonists arrived from Holland under Company auspices, they for some time eschewed Manhattan, which was esteemed mainly for the meadow grass grazed upon by the Company's sheep. Some of the island's acreage had been tilled for grain, and colonial hunters flushed out innumerable delicious rabbits (the progeny of Dutch rabbits set free on the island by farsighted sailors earlier in the century). But Manhattan was not seen as a place where a person would actually want to live.

All this changed in 1626, when Daniel van Crieckenbeeck, the commander up the Hudson River at Fort Orange, violated a Dutch promise of neutrality by joining a Mohican war party's raid against their Mohawk enemies. The captain lost his life at the hands of the Mohawk, as did three of his soldiers, including one who was served up "well roasted" at a Mohawk victory banquet. All now agreed: It seemed to have a good deal of merit, the idea of summoning the uneasy Fort Orange settlers and other far-flung colonists to live within one fortified compound as far as possible from the volatile frontier. Nut Island still would be used as a staging ground for merchant vessels. But, from 1626 forward, the island of Manhattan would host the colony's commercial capital. With a population of not quite three hundred souls, the cluster of huts at the foot of the island was the second-largest settlement in North America, after New Plimouth in Massachusetts.

By 1659, after a little more than three decades, New Amsterdam was no longer a bare-dirt outpost. But arriving immigrants found that it was still a slip of a town in the embrace of a danger-filled wilderness, ever on the verge of being overwhelmed by its natural surroundings.

Manhattan had the shape of a herring, originally a rather narrow herring, its perimeter having not yet been bulked out by the landfill that would come to comprise one-third of the lower island. The herring's pointed head faced south as if it would swim out toward the Atlantic Ocean. The settlement of New Amsterdam stood square on the fish's head, with streets laid out over a fraction of the island's twenty-two square miles of rock, lakes, streams, and forest. In the summer of 1659 a nearly unbroken carpet of green stretched across all of Manhattan and the land masses that encircled it— mainland New York, New Jersey, Long Island, and Staten Island— punctuated by a handful of farms across the wider landscape and the few scratched-out streets at the island's base. Europe had been here half a century but had left barely a footprint. On a mild summer day, an energetic woman could traverse New Amsterdam in an afternoon, even on ocean-weary legs that felt as pliable as the sinew used to tie up shop packages.

Thanks to a rare 1660 map of New Amsterdam, it is possible to locate the spot where Margaret disembarked and to trace her probable route into town. New Netherland surveyor-general Jacques Cortelyou completed the map on the orders of his employer, the West India Company, whose directors wanted a precise accounting of every feature of the town they controlled.

And they got it. In black ink on vellum, the surveyor's bird's-eye view shows the settlement in eye-popping detail. The map reveals every structure on the lower island (there are 342), every *boewerie,* every pasture. So precise was the surveyor's hand that even the tiny bucket suspended over one backyard well is pictured, as clearly as if we were peering at it from just over the hedge. Known as the Castello Plan after a villa near Florence that housed the map when its Medici-family owner acquired it in 1667 (hand-drawn maps of harbors and towns having intense collectible cachet at the time), Cortelyou's work holds the distinction of being the very first street map of urban New York. (From Amsterdam, the directors offered this response after viewing it: *Too much property given over to unproductive pleasure gardens. Allocate smaller lots!* For them, the commercial imperative ranked supreme.)

A census taken at the same time as the map's creation described many of the town's occupants and buildings. At Margaret's arrival, New Amsterdam had a population of about two thousand and boasted one hospital, one bakeshop, one gristmill, one midwife, one church . . . and twenty-one taverns. Many of the cramped and smoky tap houses huddled close by the harbor along a section of Pearl Street called the Strand. There, shipwrights plied their craft, restoring weather-beaten vessels after the bruising voyages from abroad. Walking west on Pearl Street as it traced the water's edge before cinching the narrow tip of the island, Margaret would quickly reach the heart of the town. A newcomer could reliably find a mason, cooper, or glazier here. The blocks radiating from the Strand housed many of the town's mechanics, laborers, and crafts-men. Carpenters, hatters, and tailors operated out of home work-shops in the area. Representatives of the seaborne trades—sailors, gunners, pilots, and supercargoes—also lived shoulder to shoulder in this crowded harbor front.

In the weeks to come, Margaret would often walk down to Pearl Street just to see what was going on. She would watch the stevedores hauling containers out of ships, and other workers heav-ing them onto wagons drawn by oxen, which lurched continually through the rutted streets to drop off merchandise at warehouses, workshops, stores, and tapsters. (Benjamin Franklin would later comment on the rough-surfaced streets of New York, saying you could tell a New Yorker by his gait on Philadelphia's slick paved streets, "like a parrot upon a mahogany table.") For some deliveries a hoisting wheel on the second story pulled merchandise up the front of the building using a cable and a heavy hook.

At the first bridge over the canal, the broad one at Stone Street, traders in full-skirted coats and wide-brimmed hats regularly gathered to scope out the action in the harbor and analyze the business of the day. The traders would not pay much attention to Margaret anytime soon. She was still a junior factor, concerned mainly with pressing buyers to make good on their payments for the salad oil in Valck's barrels. Oil was critical to the housewives' daily provisioning, though generally unavailable in New Netherland. That demand allowed her to negotiate a good price for each imported barrel. It was the same with pins. Everybody used straight pins to hold their clothing together in an era before zippers, with the possible exception of those who could afford more ornate and costly buttons. Yet trading salad oil and straight pins a few casks at a time could not compare in remuneration or status with selling tobacco, hogsheads of wheat berries by the hundred, or mast-length oak logs. It would only be a matter of time before Margaret would be negotiating those deals, too.

*T*he persistent clang and hum that emanated from Manhattan's maritime district announced both the settlement's ambitions and its Dutch origins, for behind this New World Lilliput fell the shadow of a very particular European giant. By the middle of the seventeenth century, the Dutch Republic (also known as the United Provinces of the Netherlands, the Netherlands, or, most commonly, eventually as Holland) had become the world's foremost seafaring nation. A diminutive superpower, Holland occupied an area only slightly larger than the states of Rhode Island, Connecticut, and Massachusetts combined, and was inhabited by fewer than two million people. By contrast, France had at least sixteen million sub-jects by 1600 and the population of England had swelled to five and a half million by the 1650s.

Despite the Dutch Republic's slight physical presence, its fortunes had become the envy of the world. The Dutch, wrote English novelist Daniel Defoe, were "the factors and Brokers of Europe. They *buy* to *sell* again, *take* in to *send* out, and the greatest Part of their vast Commerce consists in being supply'd from All parts of the World, that they may supply All the World again." The nation's thriving economy funded a formidable military and drove a heady cultural climate in which art, science, literature, and phi-losophy flourished. The list of seventeenth-century achievers starts with the philosopher Spinoza and the legal genius Hugo de Groot, the painters Rembrandt and Vermeer, and goes on and on. Com-merce, military might, and intellectual ferment fused in the century-long globe-conquering surge of energy that would gain lasting fame as the Dutch Golden Age.

In 1648 Holland concluded the Eighty Years' War, its hard-fought battle for independence from Spanish domination and reli-gious freedom for its mainly Protestant citizens. The United Provinces of the Netherlands emerged as the only republic amid the monarchies of Europe and was governed by the States-General, a body that allowed each of seven provinces to exercise its own vote. After fighting for nearly a century to win religious freedom, the republic now displayed a fundamental measure of enlighten-ment. Tolerance of spiritual and political diversity was the era's hallmark. Amsterdam printers even published books banned else-where and smuggled them into the less-forgiving nations of Europe.

Tolerance thrived on prosperity and fed it as well. A jumble of nationalities and cultures found a welcome in Holland, particularly in its big cities, and contributed to the booming economy. That economy was built on the country's maritime industries—the Dutch boasted a commercial and fishery fleet second to none. The country had more ships than England, France, Germany, Portugal, Scotland, and Spain combined; seamen comprised fully one-tenth of the adult male population. The herring industry had exploded. Each year sailors in rotund little boats called herring-busses braved North Sea swells to haul in millions of barrels of the silvery fish.

Dutch merchants sent trade vessels all over the world in quest of new commodities. In the turbulent waters of Spitsbergen, off the northern coast of Norway, Dutch whale hunters with handheld harpoons pursued their quarry from *fluyts* (translated sometimes as "fly-boats," sometimes as "flute-ships"), revolutionary vessels not half the size of the hundred-ton creatures themselves. While the luminescent oil from a bowhead whale was a jackpot, its baleen could cover the cost of an entire whaling expedition. The flexible, bony material was used to make a wealth of everyday products—skirt hoops, chair caning, and buggy whips, as well as the sexier whalebone corsets. The country's three-hundred-ship fleet dominated the seventeenth-century whaling industry and its investors swam in gold.

Whaling was dangerous and left a legion of maimed sailors to haunt the country's coastal villages, but Dutch merchants never shrank from a strong commercial prospect. On arduous sea voyages to the exotic Russian ports of Archangel and Murman, factors from the Netherlands exchanged cloth, wine, and precious metals for lumber, cod, salmon, rope, and saltpeter. In 1602 the Dutch East India Company entered trade with India, Japan, Ceylon, Buton, and Java, braving shipwrecks and pirate raids to obtain the exotic spices, rich fabrics, tea, and coffee prized by Europe's aristocrats. Dutch merchants also cornered the market in Italian marble, Hungarian copper, and what then seemed to them just another product, African slaves, as well as in salt, the essential food preservative of the day and a crucial ingredient in the herring trade. Dutch fleets descended upon the paradisiacal islands of the Caribbean with their vast salt pans. (A marble frieze atop the Royal Palace on Amsterdam's

Dam square captured the time's trade-besotted mentality: In it, a mercantile goddess known as the Amsterdam Virgin was depicted gathering to her classically draped form a cornucopia of all the world's products.)

The merchants of the Netherlands seemed to have an unerring sense of the right product at the right time. Whether they erected countinghouse fortresses on the western coast of Africa or the eastern shore of Manhattan, the master plan was always the same: find and develop lucrative commodities that would sell like waffles on fair day. The Dutch Republic minted trade opportunities in every corner of the world: sugarcane from Barbados. Mexican silver, heavy and pure. Japanese silk. Málaga pepper. The irresistible midnight richness of indigo. Delicate Bandanese nutmeg, the culinary essential of all Europe, so precious that New Amsterdam housewives kept it under lock and key. Horses bred from wild island stock on Curaçao, the preferred choice for continental equestrians with bulging pocketbooks.

When rival nations Spain, Portugal, Italy, England, and France looked to the breadth of Holland's resources, they could not help but be envious. Each country had a fleet dispersed across the high seas, intent on cornering markets and snatching territories—or at least ships—from the others. The European countries had battled, formally or informally, for centuries, since they first started to solve their problems of overpopulation and dwindling resources (and feed the never-sated hunger for exotic new goods) by sending explorers to discover new lands and new markets. As the seventeenth century progressed, Holland consistently bested its archrivals. The nation reinforced that superiority when it granted trade charters to both the West India Company and the East India Company— groups of shareholders who funded trade voyages and new settlements in exchange for profits from those explorations. It was on the payroll of the West India Company that Henry Hudson first sailed up the river that would later bear his name, and through his voyage that the Dutch claimed New Netherland as their own.

From their comfortable perch in Amsterdam, the directors who controlled the West India Company wagered that their new territory would produce storehouses brimming with riches. Few of the directors would ever set foot on New Netherland's shores to witness their venture firsthand. Personal risk seemed unnecessary as long as they could persuade capable men and women to leave their homes and settle the new lands. At the middle of the seventeenth century, the well-nourished directors were relying upon the hunger of colonists such as Margaret Hardenbroeck and her new neighbors to deliver up whatever commercial equivalent of whale oil could be discovered in the American wilderness.

$\mathcal{N}$ew Amsterdam's landmark City Tavern opened its doors in 1643. The wooden wharf outside jutted a short way into the East River, allowing patrons the convenience of coming straight off the water to their favorite (and at that time only) watering hole. The three-story tavern offered a place to congregate as a community during those long nights when just a few hundred colonists huddled against an unknown frontier. By the time New Amsterdam received its official city charter in 1653 and City Tavern (Stadt Herbergh) became City Hall (Stadt Huys), there was no shortage of places to congregate on Manhattan.

Taverns kept opening, especially along the Strand, where the pong of spilled beer and stale smoke drifted out to the street through split Dutch doors. Inside these public rec rooms, men and women rented cheap tobacco pipes (archaeologists would later unearth masses of smashed clay discards in lower Manhattan) and swigged home-fermented beer or imported "Hollands," pale yellow gin in a crockery bottle. They also consumed hard cider, a beverage staple produced from the fruit trees grown from apple seeds packed in the first Manhattan-bound ships. Illiterate drinkers identified the type of alcohol they wanted by the shape of the bottle. Gin, for example, always came in a square vessel. The inevitable brawls provided the evening's entertainment. Families came to taverns for meals, which makes sense when you consider that every colonial toddler consumed beer at mealtimes, at the very least.

The tenor of Pearl Street changed as the thoroughfare wound west. A pedestrian could browse shops such as the tasteful establishment run by mercer Simon Jansen Romaine at lower Broad Street, where he sold fine silk and lace to those who could afford such wardrobe enhancements. His stockings were so coveted that in 1661 a duo of teenagers lifted half a dozen pairs from Romaine's shop. The instigator found herself banished from the colony for eight years.

Many of Manhattan's prominent families made their homes on the streets just beyond the harbor in the shadow of Fort Amsterdam, hard by the adjacent public marketplace, Het Markvelt. The dwelling of Hans Kierstede, New Amsterdam's first physician, anchored a prominent corner lot. Records show that one patient gave Kierstede a beaver skin for stitching up a stab wound, a typical payment for the time and place. Kierstede's wife, Sara, was a renowned interpreter who would later facilitate one of the most crucial peace treaties of early New York, and a counselor to the Hackensack Indians. The tribe's sachem repaid Sara Kierstede's guidance with 2,260 acres of prime Bergen County land, making her one of the first European landowners in New Jersey.

The Kierstedes and their ten children had an assortment of luminaries for neighbors. Their yard bordered the three adjoining lots of Cornelis Steenwyck, who had built his illustrious career on public service and on shipping salt, tobacco, and enslaved Africans; in less than a decade he would govern New York City. Johannes de Decker, a prominent attorney and supreme council member, resided just down the street. Poet Jacob Steendam, the colony's foremost literary light, lived a block away on a somewhat less flashy property surrounded by fruit trees. Steendam was the author of *The Complaint of New Amsterdam to Her Mother,* an entry in the then-hot genre of Poetry as Broadside. His mildly pornographic rant targeted the dearth of funds, immigrants, and general attention allocated the Dutch colony. (A sample stanza: "When I thus began to grow, / No more care did they bestow. / Yet my breasts are full and neat, / And my hips are firmly set.") The poet would eventually disappear into the comparably full and neat bosom of Dutch Java.

Almost all the island's residents congregated here at its southern end, on the water, well below what later would become Wall Street. Pearl, Broad, Beaver, and Whitehall streets had a reputation for successful residents and comfortable properties. But the most prestigious thoroughfare in New Amsterdam was the Road, the nickname assigned to Stone Street after it received its armor of upper-Manhattan-mined cobbles, or "round stone." The colony's genteel inhabitants funded the improvement. The most famous of them, Oloff van Cortlandt, had come to New Netherland from Wyk-by-Duuerstede, just outside Utrecht, as a Company mercenary before amassing his fortune in land and trade.

Margaret did not personally know Van Cortlandt or his family—not yet—but his affluence was obvious to any passerby. In New Netherland wealth had its most clear manifestation in property. Van Cortlandt's Stone Street compound included a substantial home, four adjoining lots, assorted storehouses, and a brewery whose brisk business kept his household flush with silver plate. He even maintained a private well.

Van Cortlandt's residence, like those of many wealthy burghers around town, was faced with red Holland brick. The house had small casement windows with actual glass panes—an especially prized feature given a dearth of glass artisans in the colony. (Less-affluent homeowners would fill their window frames with oiled paper, stretched deerskin, or shaved sheets of horn. Wooden shutters alone protected many residents from the chill of winter or the mosquitoes that bred across the island's swamps and streams, and stamped inhabitants' faces with pink bites all summer.) A brass weathercock topped the tiled roof. The glossy black front door was split horizontally at waist height, allowing the top half to swing open. This let in air and light, while keeping children indoors, out of trouble.

Even along this best street, hogs rooted in the muck. Though prized for their meat, swine on the loose had a reputation for snapping at children, who would run behind the animals and flick their backs with switches. Owners had been warned to keep their animals penned. But even a wooden yoke could not always stop a hog, an animal with the wiles of a human toddler, from squeezing through fence gaps. Hog corpses littered the town, strewn beside the picturesque water beeches, locust trees, lime trees, and elms in residential lanes. The council had passed ordinances requiring residents to dispose of the remains, but it was not until Peter Stuyvesant, the colony's fourth director-general, witnessed pigs trampling the sods being used to repair his fort that he ordered a cadre of slaves to cart away the carcasses and his soldiers to shoot any further strays on sight.

The residences at the west end of Stone Street stood across from Fort Amsterdam. At just over three decades old, the fort was the most ancient and substantial structure on the island. Its shoreline position commanded unrivaled naval access to the Hudson and East rivers. If Dutch Manhattan ever suffered attack, such a well-situated bastion would guarantee its safety. (Although the building's footprint—occupied later by the beaux arts U.S. Customs House and then by the Museum of the American Indian—would remain essentially the same as it was in Dutch times, its outlook would in the future be dramatically different. Lawns, memorials, and walking paths would one day lie atop landfill that encompassed chunks of the original fort, the boulders of Capske Point, and very likely the bones of seventeenth-century sea lions that sunned themselves amid the crashing whitecaps.) The gates of the fort opened to reveal a latticed guardhouse, soldiers' barracks, the colony countinghouse and prison, and a comfortable house built for the director-general. The fortress also housed the all-important Dutch Reformed Church, its sloping, pantiled rooftop rising above the fort's earthen walls. The church's secure location offered a semblance of protection to these frontier dwellers. Inside, a painted ship's figurehead carved with the likeness of St. Nicholas stood on permanent display—salvaged, according to legend, from the bowsprit of the first vessel that carried immigrants from Holland to Manhattan. For settlers, attendance at services was a central ritual of the week.

The Jews, Catholics, Quakers, and Lutherans of New Amsterdam, in fact any denomination other than Dutch Reformed, were prohibited from building a house of worship. The city council, however, allowed worshippers of every faith to observe their beliefs "in all quietness . . . within their houses." This official tolerance was derived less from an ideal of religious freedom for all than from the West India Company's determination to populate this land, even if that meant accepting immigrants with divergent views.

As Margaret walked past the fort she passed a group of uniformed soldiers guarding a construction site, where slaves and a trio of pricey Holland stonemasons were refacing the ten-foot bulwarks. Long-term colonists were used to seeing slaves toiling on public works projects; Africans were shipped to New Netherland as early as 1626. But like most new arrivals from Patria, which did not tolerate slavery on its home turf but only in its colonies, Margaret must have found the sight of black workers laying stone at the fort somewhat jarring.

The colony gallows standing on the rock-strewn beach below was more familiar. One early artist, depicting Manhattan Island from the vantage of Nut Island, supplied labels for his drawing's landmarks; beside the slender scaffold's crossbeam he penned the Dutch legend "t'Gerecht," which translates as "court" or "justice." That particular court was not much in session during the first three decades of the colony's existence. The pillory, iron, and post were used more often to make a public example of wrongdoers, also providing a sort of public theater for the island's residents. Curiosity-seekers often went out of their way to witness a pilloried prisoner stand with his head trapped in a board.

Colonists' manners tended to match the still-wild natural environs of Manhattan. Drinkers brawled through the town's streets. Mooning was the insult of choice, even among women, despite lawmakers' efforts to punish at least one offender with deportation to Holland after she, "in presence of a respectable company . . . hoisted her petticoats up to her back, and showed them her arse." Individuals who showed verbal disrespect, on the other hand, either through slandering another person or blaspheming the Lord, tended to receive their punishment by pillory, as did street-corner beggars and culprits who engaged in "coin-clipping," "tree-polling," or "dice-cogging." One settler found himself set in the pillory for the crime of delivering false dinner invitations. Poetic justice often seemed to dictate penal policy: Colony records describe a man who, having stolen cabbages, was made to stand in the pillory for hours balancing cabbage leaves atop his head; likewise a dishonest baker, his head covered with a pile of bread dough. A narrative of a slightly later period describes the treatment given one Dr. Bastwick: "His weeping wife stood on a stool and kissed his poor pilloried face, and when his ears were cut off she placed them in a clean handkerchief and took them away, with emotions unspeakable and undying love." Men's fashionably full wigs had an ulterior function in these days of premature ear loss.

Some of the most intense punishment came under the heading of military discipline. An apparatus called the timber mare stood between Pearl Street and the fort. It was a simple sawhorse, but its straight, narrow horizontal beam hovered twelve feet off the ground. The soldier being punished straddled the beam with hands tied behind him, a weight tied to each foot. For stealing chickens, one Dutch soldier rode the wooden horse for three days, with a fifty-pound chunk of iron dragging down each foot. Another, punished as a drunk, rode the horse holding up an empty scabbard in one hand and a pitcher in the other.

More than anything, these spectacles served to contain the potential for evil, to beat it into submission and hold it at bay. A wrongdoer compelled to ride the rail around the town, carried by a team of upstanding men, would sear a lesson into the consciousness of every person who witnessed his agonized passing. That one soldier rode the horse might seem unlikely to keep other men from committing horse-worthy crimes. But the punishments at least created an idea of order in this rough, beer-swilling, moon-ready town.

$\mathcal{W}$edged between Fort Amsterdam and Capske Point, the residential fortress known as Stuyvesant's Great House turned its frost-white face toward the Strand to greet every ship entering the East River anchorage, like a billboard announcing the health and wealth of Manhattan.

It also reflected the energy of its namesake. Peter Stuyvesant, a West India Company career officer, had been installed as the colony's director-general in 1647, twelve years before Margaret arrived. Without ado, the hard-charging, thirty-seven-year-old Stuyvesant tackled the civic crisis wrought by his incompetent predecessor, Willem Kieft, whose insensitive approach to relations with the natives had doomed the settlement to a prolonged episode of guerrilla warfare.

Before the Company hired Stuyvesant, the town's population had been reduced from two thousand to about seven hundred. Many settlers had died; many more had fled back to the Netherlands or to other colonies. No police force existed to protect the populace. Kieft had so antagonized Manhattanites that they refused to contribute to the upkeep of public buildings, the traditional means of funding capital improvements. The result: a crumbling gristmill and a fort that lay "like a mole hill or ruin." The soldiers of the garrison were a drunken, brawling, insubordinate crew; a firing squad had to execute one mutinous recruit. The women of New Amsterdam ran wild. One found herself reprimanded after she, "notwithstanding her husband's presence, fumbled at the front of the breeches" of most of the men at a party and another "measured the male members of three soldiers on a broomstick."

Soon into his tenure, Stuyvesant took action on all counts, which did not earn him instant popularity. He forbade the serving of alcohol after the town bell struck nine at night or before two on Sunday. He reinforced already-standing laws against selling alcohol or arms to Indians. He imposed speed limits on wagons and carts. He banned butchers from tossing their offal out the front door of their shops. The director-general even literally straightened the streets.

Residents bristled at some of the regulations passed under Stuyvesant, taking personal offense at such measures as the clampdown on alehouses and a new prohibition against shooting partridges and other game within city limits. Who among the celebratory Dutch could endorse Stuyvesant's strictures regarding drums, maypoles, and intoxicants on New Year's Day and May Day? Colonists also balked upon hearing it was no longer acceptable to engage in "street broils or quarrels, much less beating or striking one another"—and that they would be punished with a penalty of twenty-five guilders "for a simple blow of a fist," and four times as much if blood flowed.

But the complaints did not dissuade the autocratic Stuyvesant. When summoned to The Hague in 1650 to report upon the condition of the colony, he offered a classic Stuyvesant pronouncement: "I shall do as I please."

And so he did, although he also assembled a council of nine men to advise him, a democratic first in New Netherland. Under Stuyvesant the town officially incorporated as a city in 1653. New Amsterdam now had its own *schout* (sheriff), *schepens* (aldermen), and burgomasters (justices), who formed the official legislative council. In 1657 he established the *borger-recht,* the burgherright, which meant that only individuals who kept "fire and light," that is, a permanent residence in the city, had the right to conduct business there. The restriction drew praise from New Netherlanders increasingly enraged by visitors they referred to as "Scotch factors and merchants"—no matter that they hailed from all over the globe, not just Scotland—who capitalized on the fur trade without contributing to the local economy.

As a career soldier Stuyvesant considered the settlement's defenses his most pressing challenge. On the parapet of Fort Amsterdam, above the earthen walls reinforced with lime and stone, sixteen brass cannons awaited enemies both potential and all too real. To the north, the fortress's guns pointed toward the fur-rich French colony of Canada, one of New Netherland's arch competitors, whose genius in acquiring pelts was enhanced by political alliances with mercenary Indian bands.

To the northeast, the creeping colonies of New England presented a constant headache, and disputes with England over the borders between settlements had steadily worsened. It further complicated matters that British families had increasingly installed themselves in New Amsterdam on some of the best lots in town, as respectable members of the community. Trade with the English had become integral to the business of some of the colony's most powerful citizens. Even the Dutch Reformed church within the fort had been built by Connecticut contractors, John and Richard Ogden of Stamford. Still, despite the individual ties that bound English and Dutch merchants, the larger question persisted of who could hold the land each nation claimed. A battle was inevitable. Without sufficient troops the Dutch could not physically defend their land, so Stuyvesant relied upon diplomacy to keep a grip on New Amsterdam and the rest of the settlements in the Hudson Valley.

England, not surprisingly, wanted the Dutch out of North America altogether. In 1653 the British had assembled a fleet of warships in Boston, ready to descend upon New Amsterdam. In a panic at the prospect of invasion, Stuyvesant commissioned a defensive palisade of cedar pilings at the northern limit of the city, a wall that would stretch across the island from the East River to the Hudson. To build the wall quickly, Stuyvesant commanded the entire populace—soldiers and slaves, along with burghers, farmers, blacksmiths, mechanics, merchants, sailors, women, and children—to undertake the grunt work. Shoulder to shoulder, Manhattanites dug four-foot-deep postholes and hoisted twelve-foot-long, sixteen-inch-diameter sharpened logs.

Het Cingle, the Wall, gave both a sense of protection and a geographical identity to New Amsterdam. It helped define the body of the city and even its soul. The wealthier colonists contributed funds toward its construction, and almost everyone furnished some sweat. (True to the complex politics of the time, the lumber for Het Cingle had been delivered by English entrepreneur Thomas Hall, and at a decent profit.) The Wall would outlive Dutch control of the colony. Only slightly disemboweled by citizens for firewood during lean times, it stood tall until 1699, and its shadow from end to end would be reflected centuries later as Wall Street.

To fend off an invasion, the Dutch ultimately agreed to a slate of compromises, allowing the English to settle the towns of Gravesend, Hempstead, Newtown, Flushing, and Jamaica in still-isolated eastern Long Island, which the Dutch had claimed for their own. England already had seized most of Connecticut, including the major settlement of Greenwich, which loomed only eighteen miles from New Amsterdam. The negotiations headed off a showdown on Manhattan, at least for the time being.

*T*he harbor breeze mingled with the perfume of gardens and shade trees as Margaret traversed Pearl Street and crossed the first bridge over the city's new Heerengracht Canal. The neighborhood surrounding the canal, named after one of Amsterdam's most important waterways, offered a pleasant place to meet, walk, and smoke. Colonists in America as elsewhere were quick to replicate the gamut of features that they loved in their homeland, including a marketplace, a windmill, open quays, and the all-important canal, creating a landscape that resembled nothing so much as a miniaturized Old Amsterdam. The canal that materialized at the east end of the Road originated as a wet ditch, a sluggish creek known as Blommaert's Vly that cut into the island's eastern shore and filled like a bathtub with seawater at high tide. In 1657 work began on facing the inlet's sides with timber and lengthening it almost to the Wall. Now boats laden with wood or oysters or vegetables could cruise right up the Heerengracht to unload their goods.

It was here that Margaret finally found some sign of her quarry in the form of a house on the first block of the canal, the home of Hendrick van der Vin, the accountant who still owed her cousin twenty-four beaver pelts. Van der Vin had acquired his residence at the corner of Broad and Pearl at public auction after its previous owners, a young married couple, were killed by Indians. Its location was unbeatable. Owning this particular lot meant Van der Vin had only to step out his front door to reach the base of the bridge, which gave him a useful proximity to the spot where New Amsterdam's premier burghers gathered to cement their deals. That Van der Vin was not himself a trader by profession but an accountant spoke volumes about the economic climate of New Netherland. Just as in Holland, even journeymen and servants—even accountants—put together deals, buying and selling to the extent they were able. Trade in New Netherland was not just a business, it was a way of life.

So was the tendency to default on debts. Judging from the civil court records of New Amsterdam, awaiting payment from delinquent debtors was a common state of affairs—and a source of constant rancor. Creditors tended toward lenience, perhaps because they managed their own affairs in exactly the same way.

Margaret was not so sanguine when she found her debtor not at home. But short of opening Van der Vin's house to investigate whether those packs of furs were stashed somewhere or tapping her foot on the grit-surfaced lane until he appeared, Margaret could do nothing. For now she would have to leave.

On September 19, 1659, Margaret dispatched her report to Amsterdam.

"Cousin Wolter Valck," she wrote, "I inform you herewith that I could not get payment from Hendrick Jansz. Van der Vin. After your return, he will pay you the interest due on it and he will not fail to pay the total sum."

Her bravado matched that of any junior employee justifying a questionable performance to a demanding boss. But her letter's closing sentence could have stood as her lifelong motto.

"I did my best."

3

Wild Diamonds

*W*hen Margaret landed at New Netherland, her destination's rugged landscape would have struck her as extraordinary, nearly as mind-boggling as the brick-and-mortar metropolis of Amsterdam would have appeared to the Indians who still made their seasonal encampments on Manhattan.

All over Holland, perfectly level farmland and neat little villages stretched beneath wide horizons that framed breathtaking cloud configurations. The country's ancient port cities featured tightly packed stone buildings set in a web of canals teeming with ships. Water touched everything, whether as ocean, lake, river, or marsh, though it had mainly been corralled, contained to whatever extent possible behind dikes and dams.

Manhattan could not have been more different: Nature remained uncontrolled. Streams flowed from the island's highest points to the East, Hudson, and Harlem rivers. One stream, called Montagne's Brook, meandered from what would become the intersection of Eighth Avenue and 120th Street across the island to the East River at 107th Street. The Saw Kill ran from the future

Park Avenue into the East River at 75th Street. Through the neighborhood we know as Greenwich Village, then a native encampment called Sapponikan, wound little Minetta Brook, the only vestige of which now is a trickle through modern basements. Ponds abounded, too, with evocative names: Buttermilk Pond lay just across a winding, rustic road (the Bowery) from Sweetmilk Pond. The ice-cold, seventy-foot-deep Freshwater Pond, Manhattan's longtime source for uncontaminated drinking water, occupied ten acres at the heart of the future Chinatown.

The climate and soil of Manhattan collaborated to produce a surfeit of vegetation. Grapevines so dominated the hills beyond New Amsterdam's cluster of houses that horses tripped in the tangles, reported Adriaen van der Donck, a lawyer who served as the first sheriff for the Dutch settlement of Rensselaerwijk, 150 miles up the Hudson from Manhattan. A voyager named David de Vries described the fruit as "large as the joints of the fingers," and its dense sugar and fleshy weight gave it the name pork grape. Pressed for wine, this specimen confirmed its wild nature, rendering a juice so dark Van der Donck commented it "resembled dragon's blood more than wine."

Colonists shared their living quarters with wild animals, not necessarily intentionally. Bears invaded carefully tended orchards. A colonist named Charles Woolley recorded the time he "followed a Bear from Tree to Tree, upon which he could swarm like a Cat." Some inhabitants lived like beasts themselves. Along De Heere Straet—also called the Public Wagon Road, the Great Highway, the Common Highway, or simply the Broad Way—settlers of lesser means actually made their homes in square pits dug in the ground, smoky cellars lined with bark and roofed with sod. A Manhattan summer brought a soup of humidity and temperatures topping one hundred degrees. In winter the cold was bitter enough to freeze eels in a bucket of water. It was so cold that people brought their chickens inside to warm them by the fire.

Nature took on vast dimensions in this new world. Some of its creations puzzled newcomers. The hummingbird, for instance, had no counterpart in Europe. Curious visitors such as Swedish botanist Peter Kalm surmised that the creature must be a cross between bird and insect, hovering in midair and returning again and again to "suck nourishment . . . like the bees" from blooms heavy with nectar. The little marvel had so delicate a constitution, he noted, it would perish if caged as a pet the way people kept robins or catbirds. Adriaen van der Donck favored a creative use for these beautiful creatures, which were too "tender" to make good pets: "We . . . prepare and preserve them between paper," he wrote, "and dry them in the sun, and send them as presents to our friends."

Other animals also provoked wonder, not only for their novelty but also for their magnitude. The outsized bounty of New Netherland had become the stuff of legend by the time Margaret arrived. Settlers regularly harvested three-foot and sometimes six-foot lobsters from the floor of the East River harbor. Monstrous pelicans stormed the horizon over the velvet ribbon of Long Island. Whales heaved themselves along the Atlantic coast "and . . . the bay, where they frequently ground on the shoals and bars," wrote Van der Donck.

The Hudson teamed with schools of salmon, drum, carp, snook, pike, flounder, suckers, thickheads, eels, and sunfish. Van der Donck noted flocks of pigeons that "resemble[d] the clouds in the heavens, and obstruct[ed] the rays of the sun." In the 1640s De Vries observed one hunter in the commander's orchard at Fort Amsterdam pull down eighty-four pigeons. Just offshore, crabs crawled eagerly into fishing nets, and people took for granted oysters the size of a man's hand. All around the island, the Indians had piled discarded oyster shells a dozen feet deep, evidence of years of feasting—Pearl Street received its name from the shell middens in its vicinity. The supply showed no signs of diminishing. Water terrapin, a delicacy in Europe, abounded. So did the luscious striped bass, "a delicate, fine, fat fish," according to New England chronicler William Wood in 1634, "having a bone in his head which contains a saucerful of marrow, sweet and good, pleasant to the palate and wholesome to the stomach." A man could catch a dozen and a half bass in three hours, using a chunk of lobster as bait.

Colonists praised the sights and smells of North America. They professed amazement at the sheer "sweetness of the Air," as English pamphleteer Daniel Denton phrased it. "Much," quoth the Reverend John Miller, "like that of the best parts of France." The wind felt fresher than that of any country in Europe. The flaming reds and golds of a New Netherland autumn had no parallel in Holland. The region west of the Hudson, wrote Cornelis van Tienhoven, the colony's fiscal under Stuyvesant, was "the handsomest and pleasantest country that man can behold." Winter storms produced the gorgeous novelty of "trees silvered over with sleet." Naturalist Kalm suggested the primordial flavor of Manhattan with an observation about the island's population of tree toads, which "frequently make such a noise that it is difficult for a person to make himself heard."

Margaret's passion lay not in New Netherland's beauty so much as in its bounty. She came to trade. She likely looked at the Hudson River less with a poet's eye, or a naturalist's, than that of a commercial novice who was there to make good. She was shrewd enough to grasp that she needed the Hudson for her success in the long term. The massive waterway offered the one practical means of entering this vast, unmapped country. The river would be her door. Now she needed its key.

*T*he country beyond the reach of Manhattan's streets, canal, and great river was accessible by the native footpath that ran from Fort Amsterdam to the very top of the island and even beyond. Colonists already had transformed the path into the road known as the

Broad Way. It began near an ancient canoe landing on the shore of the Hudson and moved in one straight shot up the west side of Manhattan as the island's principal north-south thoroughfare. At the Wall, the road veered east, then continued north. The trail had been the Indians' main passage on Manhattan. Now natives shared the thoroughfare with colonists: Indians hefted canoes above their heads alongside Dutch farmers heading toward the heart of town with loads of maize and root vegetables.

Above the palisade, horses could be seen racing through undeveloped bluffs and meadows. Many were descendants of thoroughbreds that busted away from their masters after being shipped to America from the Dutch colony of Curaçao, and they continued to multiply in the wild. Farther along, the native settlement of Konaande Kongh sat tucked beneath an impenetrable line of soaring boulders at the northern end of the future Central Park. The Broad Way was the only trail through this natural barrier, though nineteenth-century blasting later would alter the topography. The path continued through the new village of Haarlem, home to three dozen citizens who built houses in 1658 on the site of earlier failed settlements. The West Indian Company specifically promised that the village would promote agriculture and provide "recreation and amusement" for the city.

Haarlem was the last European outpost on the northern end of the Manhattan landmass, But intrepid colonists could continue along the trail into the woods to the landmark known as the Wading Place, on the northern lip of the island. In these elevated reaches, full of snug rock shelters and game-filled glades, it would be possible even as late as Margaret's arrival to observe thriving Indian villages, the last holdouts in the epic native effort to coexist on Manhattan with the men and women who had come on ships to make their lives among them.

———

*W*hen the Dutch first set out constructing their bark huts at the foot of Manhattan, various Lenape villages—more accurately seasonal encampments, because the Lenapes were a nomadic people—could be found from one end of the island to the other. Theirs was a thriving, complex society comprised of distinct village bands, counting a few hundred members each, many of whose names (the Raritans, the Canarsees, the Hackensacks, for example) still correspond with the territory they claimed as their home turf. On the west bank of the Hudson, a well-worn trail led to Sapponikan, "tobacco field," the ancient agricultural center and fishing site. In the far northern precincts of the island, Papperinemin had long been established on Spuyten Duyvil creek. Another settlement, called Naigianac or "sandlands," lay on the southeastern shore at the outcropping the Dutch named Corlears Hook, hard by a good-size freshwater stream. An irregular grid of trails, built high along the island's rocks and ridges, stretched between the various encampments. The Indians relied on the trails for swift passage over the rough terrain that earned the island its Lenape name, Manatus, "island of hills."

The Lenapes of Manhattan, Long Island, and New Jersey made regular forays along these trails and across the water to the taverns, shops, and homes of Dutch New Amsterdam. Groups of up to three hundred natives regularly came to trade and sometimes camp overnight. The white bread and sweet cakes at the bakeshops were a reliable draw: Before the advent of Europeans, the Indians had never tasted wheat flour or cane sugar. (Bakers were happy for their business, but lawmakers were quick to curtail native pastry privileges when a wheat harvest came in short.)

Colonists derided their native neighbors as *wilden,* savages, but observed them closely and sometimes even forged an intimacy. One central character in the colony, Secretary Cornelius van Tienhoven, was noted for his fluency in various Indian languages and his dalliances with native women. Garrison sergeant Nicolaes Coorn found himself brought up on charges because, according to the sheriff, he "at diverse times had Indian women and Negresses sleep entire nights with him in his bed, in the presence of all the soldiers." The Indian men kept their heads shaved with hot stones, except for a standing roach of stiff black hair, with one dangling scalp lock at the crown; they covered their nakedness with leather breechclouts and dressed skins of deer or wolf. The women wore only fringed girdles of Indian hemp, which fell well above their leggings. In the 1640s the lawyer Van der Donck pronounced the women so "fascinating" that the single men who "were connected with them" before the arrival of Dutch maidens "remain firm in their attachments." Eventually the offspring of these couples walked the streets of every settlement in New Netherland.

The Lenapes still exhibited a strong presence on Manhattan in the middle of the seventeenth century, a testimony to their culture's vitality, but a series of epidemics had already begun to thin the native population to one-tenth its original size. Smallpox, the worst plague, had arrived with explorers in the 1640s and emptied encampments all over the vast Lenape territory known as Lenapehoking. The fever, pustular rash, and blindness killed most victims in just two weeks. Sweat baths and river dunkings that were intended as therapeutic worsened symptoms. Nursing the sick spelled suicide; flight further spread contagion. One observer told of wolves that smelled death and devoured the corpses, then attacked those too weak to run. At least seven epidemics swept through the region between 1633 and 1691. For a brief period the number of Indians equaled their new European neighbors—the difference being that every week ships arrived from Europe carrying men, women, and children to replenish settlers' ranks.

At the same time, violent episodes between colonists and natives increasingly poisoned the possibility of coexistence. The devastation cut both ways. By the time Margaret arrived, it had grown patently clear that a warrior could wreak nearly as much havoc with a war club made out of a blunt stone and a stick as a Dutch soldier could with his professionally crafted musket. Soon the natives acquired guns, bartering skins for weapons. (A pelt gun, for instance, earned its trader the exact height of the weapon in stacked furs.) Settlers, though, had the inevitable advantage of numbers and of firepower. During the tenure of Willem Kieft, the director-general who preceded Stuyvesant, relations between Dutch and Lenapes had spun out of control. In a three-year period, many hundreds of Indians and colonists would die.

Kieft's successor was hardly more enlightened; Stuyvesant's self-professed mission was to cleanse the Hudson Valley of the remainder of the *wilden*. He habitually presented Indian captives as gifts to his more deserving troops. Yet no matter how harsh the enmity between colonists and natives, hostilities could not erase a fundamental interdependence, a relationship grounded in commerce. For both Indians and settlers in the Hudson Valley, trade constituted their lifeblood.

*A*long the massive waterway to the west of Manhattan Island, traders who followed in the wake of Henry Hudson's pioneering voyage anchored in the shallows of the river—their floating trading post—and only had to wait. Those who lived along the river's banks inevitably would arrive holding armfuls of skins to barter, engaging in a practice described as the "fugitive trade." Some inland Lenapes traveled a month or more to do business with the Dutch on the shores of the river they called Muhheakunnuk, the "river that flows two ways," named for the north-south tidal pull of its waters. Colonists dubbed it the North River; only later would it be known as the Hudson.

Together, the natives and the Dutch traders navigated a complicated but often amicable relationship based on the mutually beneficial traffic in goods, despite mistrust, ongoing violence, and territorial disputes. They even had a private language, a deal-making jargon that pooled European and Indian terms. From the start, the natives welcomed novelties such as iron axes, kettles, and hoes fashioned of metal, everyday commodities for Europeans. A blade of steel was a revelation.

Textiles, too, drove the business of barter. Entire towns in Europe based their economies on producing the woolens used to trade with Native Americans. (The town of Witney, England, which sold a popular kind of cloth called duffels, kept 150 looms in operation and employed three thousand townspeople to keep production high.) While Indians did not give up wearing pelts, they augmented their wardrobes with a sackclothlike linen that came off European factory looms. To them, bolts of cloth were a worthy exchange for the furs that were easy to hunt and readily replaceable. One account tells of an Indian trader exhausting his fur supply during the course of a day's barter; in his eagerness to obtain the items at hand, he ripped off his own beaver wrap to deliver as payment. By midcentury the traditional trade goods had been augmented by gunpowder, guns, and alcohol, which proved so addictive that many sachems, tribal spiritual leaders, pleaded with merchants not to barter these items with their people.

The classic commodity of the fur trade, though, was wampum, better known to the Dutch by the Algonquin term *sewan*. Belts of strung shell beads had always played a central ceremonial role for the Indians of the Northeast—as a gesture of goodwill, the Mahicans who approached Henry Hudson offered him "stropes of beads." But the beads had not served as a unit of currency until the natives began trading with Europeans. The Shinnecock, who lived on Long Island near the Atlantic, made *sewan* a cottage industry, painstakingly drilling the beads with primitive tools. Women gathered the shells of spiral-centered whelks and periwinkles to make white beads and dug the thick-shelled quahog to craft the purple beads, called black wampum, which had twice the value of the white. (The clam's scientific designation, *Mercenaria mercenaria,* would reference its value as currency.)

Once merchants recognized the value of wampum, they pushed to develop the resource, even establishing primitive "factories" at oceanside trading posts. They installed native craftsmen to perform the labor of shell-shaping with imported metal drills, rather than the stone-tipped awls that had been used for four thousand years. As with all forms of colonial currency, the value of wampum fluctuated dramatically and usually was expressed in terms of guilders, even though guilders were rarely exchanged in the colony. A man could buy a house either with wampum, kegs of beer, or a quantity of clapboards. Company soldiers received their pay in wampum. Beaver skins represented the closest thing to gold in the colony, but wampum was a close second.

Wampum was perhaps the perfect metaphor for America's bounty: a surfeit of seashells transformed with a little sweat and skill into money. The coin of the realm, it grew ubiquitous as currency for New Netherland's shopkeepers, tapsters, farmers, artisans, and housewives, especially since the colony suffered chronic shortages of gold and silver and even hides. Soon enough, the Dutch Reformed Church in New Amsterdam considered it a matter of course for its well-heeled parishioners to deposit stropes of shell beads in the collection plates on Sunday.

*I*f there was a problem in the early days of the colony, it was one of superabundance. For a decade before the formation of the West India Company, Holland's major merchants sent emissaries to the Hudson River to evaluate available resources. These early traders returned with galleons stuffed with the fruits of the land, encouraging the establishment of the colony in 1624.

In 1626 Peter Minuit purchased the island of Manhattan from the Canarsee for the equivalent of sixty guilders paid out in trade goods. Not a hefty sum, at a time when the capture of a single Spanish treasure fleet netted the Dutch fifteen million guilders, but Minuit, the third governor of the colony, nonetheless deemed the price "reasonable," and the transaction "honorable." He had received strict marching orders from the West India Company (inspired undoubtedly by the Crieckenbeeck fiasco, when four Dutch men died at the hands of the Mohawk): Any Indians inhabiting Manhattan "should not be driven away by force or threats, but should be persuaded by kind words or otherwise by giving them something, to let us live amongst them." The Dutch had learned from previous colonial experience that achieving their mercantile goals would proceed more expeditiously if settlers approached the natives with at least a veneer of respect.

That same year a ship called the *Arms of Amsterdam* had cleared the island's harbor on its return trip to the Old World. Its manifest, the earliest known for any ship departing the port, lists "many logs of oak and nut wood"—a cargo of prime shipbuilding materials sure to fetch an excellent price on the European market. Good ships were worth a fortune and made exclusively of wood. Europe's forests were depleted, making a new source for timber an utter necessity. Shipbuilders had long been bereft of their giants, their mast stock, forcing them to bolt together shorter logs to make one mast of the proper height. Merchants, therefore, earned the biggest profits from mastwood, even calculating a greater investment to harvest and transport the gargantuan pine trunks, the best of which measured more than twenty-four inches in diameter at the butt end.

In New Netherland, massive specimens stood right outside the door. Even Manhattan, relatively crowded with settlers, had a worldwide reputation for its magnificent white pines. Colonial lumberjacks described discoveries of incredible dimensions—the surfaces of some stumps, it was said, were sufficiently broad for a team of oxen to stand upon. Oaks grew sixty or seventy feet high without a knot. It seemed that the supply of American hardwood would never be exhausted.

Tobacco also intrigued the settlers, who saw their English neighbors in Virginia make fortunes cultivating the crop. They found that Manhattan's fields, cleared by Indians at some point in the past, also made excellent tobacco plantations. Some colonists harvested sassafras, the native tree that was one of the first New World exports to Europe. The extracted oil of sassafras roots made the tree a near-overnight sensation, sought for candies, soap, and a tea that healed a panoply of ailments. (Sassafras was thought to be so potent that physicians would only treat plague victims while wearing "nose beaks" containing the herb.) The West India Company believed the craggy hills of New Netherland might hide precious jewels, and that its multitude of mulberry trees might nourish silkworms. Even harbor whaling offered a fantasy of remuneration, until it turned out that the pods that made their home at the mouth of the Delaware were not prolific enough to make the hunting a worthwhile investment.

Another mammalian resource would be a much more reliable earner. From the first years of the colony, animal hides stocked merchants' homebound ships. In addition to logs, the *Arms of Amsterdam* also carried 8,250 pelts on its 1626 voyage, representing about a year's worth of wheeling and dealing between Dutch traders and Indian hunters. Imagine the mechanism that must have existed, so early on in the life of the colony, to convey that number of skins from seller to buyer, the speed with which it materialized, the passion that drove its operation. There had to be a cadre of traders on overdrive moving up and down the banks of the Hudson River to negotiate deals with the native trappers, who for their part already had begun to sacrifice months away from their communities to deliver the goods. And with time the machine of trade grew only more efficient. Scholars estimate that annual exports of hides from New Netherland more than doubled between 1624 and 1635, and peaked at eighty thousand each year in the 1650s.

Dozens of species of fur animal roamed New Netherland's forests, and each type offered distinctive features that increased its market value. The white-tailed deer's tough skin, for example, made durable everyday breeches and shoes. Mink could be stitched into sleek collars and cape linings. So could the small, short-haired bobcat that menaced settlers from the branches of trees. The shiny, pitch-dark coat of the black bear was "proper for muffs," and coveted by elegant dressers, according to Dutch observer Van der Donck. Even fashionable men considered the muff a wardrobe staple equivalent to a twenty-first-century pair of leather driving gloves.

But from early in the colony's history, a single fur animal elicited the most intense European trade activity. This was a creature of such overarching importance to the Dutch that its pointy nose, mesomorphic silhouette, and frying-pan tail were faithfully rendered in the design of both the official shield of New Amsterdam and the provincial seal of the colony of New Netherland.

It was only a rodent, but the beaver had something the world was clamoring for.

Early in the seventeenth century, the Dutch had developed an entirely new use for the fur of the beaver. By actually disassembling the pelt itself and rubbing together its barbed fibers with liquid and nitrate of mercury (the chemical that made the Mad Hatter mad) master craftsmen created the finest felt in the world. It was a product with only one purpose: the manufacture of hats.

By the mid-1600s beaver fur and high-quality hats were so inseparable that felt hats came to be known simply as beavers, as in, "Please, take off your beaver and sit down." Hatters trying to economize might fold in other types of fur—rabbit was a popular addition—but none had the same flawless results. (The typical beaver's wide brim and waterproof texture also offered practicality in a time before umbrellas.) England's charismatic style-setter Charles II championed the fashion in tandem with an ultralong, black coiled periwig, making public appearances in a wide-brimmed felt chapeau, one side tacked up, finished with a snowy swooping ostrich plume. The beaver became as indispensable to the Restoration gentleman's wardrobe as the bright red high heels on his square-toed boots.

The fashion swept Europe, drifting down to the middle class, to anyone who could possibly afford to buy a beaver. Versions multiplied. Fops flaunted the cavalier mode, sometimes draping gilt braid around the crown to match the gaudy decorations on their cuffs and breeches. Puritans wore beavers, too, but usually drew the line at feathers, preferring a plain style with a tall, stiff crown, adorned if at all by a simple silver buckle. Though most women of the era wore linen caps, fine ladies also topped their outfits with beavers, generally a bonnet with a soft, slouchy crown and a generous brim. The most fashion forward took the riskier move of modeling hats identical to those of men. (Gender-bending should not have shocked anyone, though, in a time when men laced themselves into corsets just like women's and padded their calves and buttocks to enhance their sex appeal.)

Nantucket Island in Massachusetts was purchased in 1659 for thirty pounds plus two beaver hats, one for Governor Thomas Mayhew and one for his wife. Well-to-do gentlemen on either continent made hats felted of beaver part of the uniform of daily life, and rarely removed them, even at meals or in church. A gentleman only denuded his head when the mounded curls of his periwig grew too fashionably full to balance a hat, and even then he went nowhere without his hat tucked under his arm.

There was, some Dutch merchants perceived, just one problem

with the lucrative beaver business. They worried that the supply of pelts could not possibly meet the skyrocketing demand. In seventeenth-century Russia, where pelts had once been "common as dung," according to Canon Adam Bremen, the animal had been trapped out, hunted to extinction, just like in congested Europe. For fur traders, the American wilderness offered salvation.

The North American beaver, *Castor canadensis,* had populated the vast stream-cut terrain since prehistory, with the species evolving over time to reduce its girth from roughly the size of a rodeo steer—reputedly able to face off against a saber-toothed cat—to its current dimensions, closer to a standard poodle, only pudgier and lower to the ground. The natives seemed to have no trouble catching beaver, especially in the groggy depths of winter when hunters could chop open the icy skin of a pond to pull whole families from their lair at once. Many colonists figured that all the hunting in the world would never make a dent in the animal's population.

The Hollanders who settled New Netherland in the colony's infancy saw that they could prosper by satisfying the demand for pelts to make hats, a commodity they themselves coveted. For the most serious merchants, the business of buying and selling furs soon came to resemble nothing so much as trading in diamonds (another perennial Dutch passion). But even the smaller entrepreneurs hungered to acquire the furry gems of New Netherland. Their passion became the moving force behind the settlement of the territory.

It was said that a beaver hat well stiffened with gum Arabic and Flaunder's glue could bear the weight of a two-hundred-pound man. In similar fashion, the business of beaver could easily support the ambitions of a trader like Margaret. Young and untried, she yearned to get into the market, to find a place at the table of fur, just as less-moneyed investors had entered the herring trade and made something of themselves.

If nothing else, she could get a hat out of it.

4

A Wedding, a Child, and a Funeral on the Ditch

Margaret entered her first marriage with nothing but her ambition and a child in her belly. She emerged from it a widow, a mother, and the owner of a fleet of trading ships.

Dramatic changes of life station were common in New Amsterdam—indeed, they were part of the appeal for an independent person like Margaret. It was customary for a poor worker, even a servant, to advance so far in even a brief lifetime as to leave his descendants an enviable estate. Fortunes materialized swiftly. Epidemics killed suddenly. So did shipwrecks and native attacks. Everyone knew the risks, and there was no advantage to a long engagement.

The courtship of Margaret and a fifty-six-year-old widower named Pieter Rudolphus de Vries, a trader of some consequence in New Amsterdam, took place amid the six-week-long festivities of the annual autumn beast market. Every October, the beast market offered the opportunity for full-bore drinking, eating, firing guns, and coupling. For privacy, people stole away to quiet spots such as the grassy hill behind Stuyvesant's Great House, known as the Locust Trees, where they gazed over the black water of the harbor, the mass of stars reflected like a scattered handful of sugar on its surface, with the harvest moon floating above. Merrymakers partnered up in unfamiliar beds all over town and on the gritty sand beach of the East River, the prickly floors of the pine forests, and the soft grass carpets of meadow grass. These liaisons did not all end in weddings, but some did.

The first purpose of the festival was not to party but to buy and sell. For six weeks the West India Company welcomed merchants from outside Manhattan to sell their wares in the colony, a privilege that usually wasn't granted to anyone but resident burghers. The Company posted official proclamations as far away as the English colony of New Haven and the wilds of eastern Long Island inviting all farmers to bring their herds to Manhattan. The country people brought beef, along with wild and domesticated fowl— ducks, quails, pigeons, turkeys, and swans. Fishermen set out baskets of blue crabs and lobsters.

Diversions filled the entire town. The official market took place at Het Markvelt, the plaza at the side of the fort, where vendors put up booths decorated with flags and strings of flowers. Blowing trumpets, they lured buyers to their tents. Sellers hawked cattle, but also lace, gloves, furs, ribbons, watches, jewelry, and perfume. Children raced around with pinwheels.

In some ways the beast fair was a paler version of the ancient Dutch harvest festival known as *kermis*, with its fire-eaters, steeple-walkers, and greasy fair food. At the *kermis* in Holland, men wore women's clothes and women dressed as men. The nights filled with grotesque masks and unexpected noises. Freaks appeared: dwarfs, giants, bearded ladies, dried mermaids, imbeciles, dangerous convicts, the insane. *Kermis* was about weirdness, the edges of propriety, but it also was a celebration of plenty. Fairgoers fed on milk pap, buttered waffles, and the special *kermis* dish—cake iced with mustard and sugar, surrounded by biscuits, capers, and raisins. Wine flowed, especially wine stiffened with sugar. A farmer led the *kermis* ox from house to house so residents could identify the cuts of meat they would buy when the animal was slaughtered.

At New Amsterdam's beast market, target shooting rather than cross-dressing took center stage. Men fired their muskets all over town, even as people danced in the street. Card players placed bets. The tradition known as Pulling the Goose also went on, though surreptitiously; it had been banned by official edict since 1654. For the contest, a live bird's head and neck were greased with oil or soap, and it was suspended by ropes between two poles. Young men on horseback rode at full tilt toward it, and whoever seized the bird was crowned king of the fair. Kitten in a Cask, another favorite game, had players whack a sealed keg with a stick like a wooden piñata until the cat managed to escape or died trying. But the most popular pastime was *kolven,* the precursor to golf, a rowdy game that had participants hit at balls with sticks, sometimes lobbing them into passersby or smashing windows in the process.

It was amid these festivities and under the harvest moon that Margaret promised herself to her husband-to-be. On October 10, 1659, the Dutch Reformed Church of New Amsterdam published the marriage banns of Margaret Hardenbroeck and Pieter Rudolphus de Vries. Dutch law stipulated a three-week betrothal period, during which a couple's intention to marry must be published three times at church, city hall, or the location of the ceremony. By the time winter arrived, the couple would be wed.

At twenty-two, Margaret embarked on married life at an age that was typical for most Dutch women, other than the gentry, who often married in their teens. Margaret, rather than her parents, chose her husband, which also typified Dutch betrothals. It was considered ill-advised for parents to micromanage a daughter's choice of a mate, and there was no need for her parents to give their official blessing because the age of consent for brides was twenty (twenty-five for grooms).

Nor was it out of the ordinary that the bride had only resided in the colony a few short months before the wedding. Her march to the altar may not have been as swift as it sounds: Pieter made a trip three years earlier to Amsterdam, where he conducted business with Valck—a transaction Margaret would have helped execute as Valck's factor. If their courtship began on that visit—she would have been nineteen at the time, and her future husband fifty-three, a good six years older than her father—their relationship might well have continued on subsequent journeys Pieter took to the fatherland.

The only element that suggested other motives for their union was the timing. The next blessed event in their lives would come roughly eight months after their fall wedding—the birth of a baby. The Dutch Reformed Church judged both premarital sex and out-of-wedlock childbirth as sins. And so, perhaps, Margaret and Pieter joined the ranks of couples throughout history that hustled to the altar to square their couplehood with their church and give their child an unblemished start in the world. It would have been customary for Margaret to wait for her parents to arrive and witness the confirmation of her vows in church, then attend a rollicking party with all her family and friends. But the baby, of course, would not wait.

Margaret and Pieter had a chance to establish their new partnership with little distraction. After the wedding, commerce was frozen as solid as the Hudson River. From November on, Margaret could stand facing west on the rise of land in the center of the island and see a nearly solid expanse of ice, silver-white, jammed with upturned shards that creaked as they rubbed against one another. There were no more boats for months after that. No products could be shipped to Fort Orange, and no exports—no skins, wheat, or timber—could travel the many miles down to Manhattan.

As for communication, there were now just two methods for sending important dispatches. Some couriers used their Holland-honed skating skills to slide along the frozen river past the hills that rose on either side. The alternative was Indian post. Native runners carried letters on foot along winding riverside trails in numbing temperatures and snow up to their hips. In these frigid months, Manhattan's colonists could do little outdoors and little to further their income.

Margaret's first American spring brought new identities for the thawed river, a new appearance with every visit she made to the water's edge. Its surface looked almost solid some days, like chipped blue flint. Sometimes it appeared as cloudy as a knife of tarnished silver. Other days it seemed to flow like olive oil straight from the cask. Once again the Hudson carried boats to and from Fort Orange.

The free passage of ship traffic was a matter of great concern for Margaret's new husband. At fifty-six, Pieter was a relative silverback among the *kolven* players and had a reputation as one of the more prosperous men on Manhattan. Judging from his appearance in city records, Pieter probably had not arrived in New Netherland before the middle of the 1650s, by which time he already had gained some success as a trader. By 1658 he was elected *schepen,* alderman, a position for which he would have to have had the great burgherright—a status only twenty colonists were flush enough to qualify for at that time. Pieter, though, could afford the payment, with his profits from selling bricks, paper, tobacco, sugar, wine, cider, lumber, and prunes. Real estate also interested him, and by the time he met Margaret he already possessed lots to rent or sell in both New Amsterdam and New Jersey. He had illustrious family connections as well. His mother was a blood cousin of the wife of the great painter Rembrandt van Rijn, an affiliation that must have enhanced Pieter's stature in a time of widespread veneration for the fine arts.

After the couple married, Pieter sold his widower's lodgings on Winckel Street and put up a new house on the west side of Broad, on a substantial lot fronting the canal. The Manhattan street plan of 1660 shows the couple's low-slung structure, which stretched across the entire street-side face of the property, with an adjacent slant-roofed shed that in keeping with conventions of the time probably was used as a kitchen. That lean-to cooking facility gave the house a slight edge of luxury; the more typical New Netherland homestead was one single room measuring twenty-by-twenty feet, topped by a shallow garret used for storage. Most of the houses surveyed on the early map, for example, clearly fit the one-room, floor-and-a-half model. They were interspersed with just a few dozen more substantial homes and warehouses. But tight spaces and low profiles did not necessarily peg their owners as unfortunates. Even the wealthiest, most influential homeowners did not necessarily rush to add additional rooms or floors. The obdurate traditionalism of the Dutch often meant holding on to a house in its original state, even when a property owner could afford to renovate and expand, in the same way that family members might cherish and continue to use a tarnished silver plate over the generations, even when they could afford to buy as many shiny new plates as their hearts desired.

What mattered more in signifying status was the location of a house. Did the property overlook the harbor? Did it lie hard by the fort, on Het Markvelt? These were high-traffic areas, where a Hollander on his way up could flaunt his position. While Pieter enjoyed a modicum of success, the fact that he had not reached the pinnacle is evidenced by the situation of the house he owned with Margaret—on prestigious Broad Street, yes, but well up from the harbor, in the relative boondocks of the community. (The Road lay well out of reach.) In back, the property was less cultivated than the best lots in town, stretching out in an irregular triangle that sloped down to a swampy area known as Cripplebush. There was no orchard, another status symbol on Manhattan. Residents generally kept as many fruit trees as space allowed—a few, a dozen, an acreful if they had an acre. Colonists ate fresh cherries and peaches like candy and raved about the experience in their descriptions of the place. Margaret and Pieter could cultivate their land, but it would take time.

As of the early 1660s, only the best houses in the colony were built of brick, partly because only the richest homeowners could afford to hire specialists who knew the craft of bricklaying. The bricks had to be imported from Holland, usually as ballast, including both large dark burgundy *moppen* and the hard yellow type, the "clinkers," made from clay dug out of the old bed of the Rhine and named for the sound they made when dropped on the ground. The ideal for house walls was a composite of the two, with the yellow bricks forming the trim and accent, offset by dark red bricks and later, as Albany began founding brickyards, the distinctive orange-rust tones of that city's product. Sometimes, cross-bonding would produce a more flamboyant effect, such as the pattern known as "bacon layers" in which dark and pale bricks alternated. But prices were high and ship deliveries sporadic. As a trader not yet completely at the top of his game, Pieter had his cash locked up in his investments. And so the house was constructed of clapboards; the immensity of the surrounding forest provided dirt-cheap lumber. Wood siding may not have had the most elegant appearance, but one of the province's woodworkers could easily accomplish a neat and respectable dwelling.

The environs of the house on the canal were quiet, it was that far away from the heart of town. But every morning one sound broke through the tranquillity: a pack of noisy children arriving for school. Alexander Carolus Curtius, the first schoolmaster for the colony's new Latin School, taught class every day at the house the Company built for him a few paces up Broad. He had arrived on Manhattan in the same season as Margaret. Every day adolescent boys arrived late and ran wild, brawling in the classroom just as their fathers did in the taverns. But still they appeared, testimony to the value their parents placed on mastery of the Latin tongue, a skill the Dutch perceived as crucial to a civilized existence. (At least for boys, as girls could not attend the Latin School.)

Curtius did not last long. Soon after arriving from Lithuania, the schoolmaster began to petition the Company for a raise, while charging some families as much as a beaver skin—he called the payments "presents from the parents"—and devoting his spare time to practicing medicine. Less than a year after Margaret and Pieter moved into their house, the Latin School went silent. By then, the couple had their own child.

*D*uring the nine-month period known in seventeenth-century locution as the "breeding time," Margaret needed adequate garments that suited her still-active lifestyle. No maternity wardrobe existed, so Margaret adapted her usual costume for outings and received evening visitors in a flannel bed gown. Pregnant Dutch women did not retreat from the world. Margaret's peers both in the New World and Holland frequently sewed a drawstring into the waistbands of their skirts, while the more iconoclastic made the daring move to abandon whale-boned stays altogether. Some expectant mothers wore quilted waistcoats, similar to those worn by men.

There was room for improvisation because Dutch fashions had grown considerably less formal and uniform during the second half of the seventeenth century. Gone were the ultrastiff neck ruffs of lace-edged cambric, their dozens of yards of plaits painstakingly starched and pressed. The new style, less ostentatious, was known as "falling collars." The sleeves of women's dresses featured delicate ruffles to match the collars. Petticoats of lavender or scarlet—public attire rather than undergarments—were set off by the generous shirred yardage of a robe in a complementary jewel tone. A pointed bodice, decorative apron, and leather gloves completed the typical ensemble, atop buttoned or beribboned slippers whose charms rarely would be glimpsed beneath those voluminous skirts.

In keeping with the custom of the time, Margaret worked throughout her pregnancy, bartering and collecting debts. She had to continue. She was building a business with her new husband, as well as pursuing assignments for her cousin Valck. The midwife, *vroedvrouw,* meaning wise woman, had not advised any limit on Margaret's activities.

The West India Company had brought the young midwife, Trijn Jonas, from Sweden in 1630, and her wisdom came from decades of delivering the babies of energetic mothers on the Manhattan frontier. The colony's master carpenters had put up a comfortable cottage for Trijn on Pearl Street, under a large and graceful shade tree beside the fort. Every citizen knew where to find her to catch a baby that would not wait. And if the midwife was not at home, if she had gone to the fish market or to wash her linens at the brook, her conscientious neighbor, the poet Jacob Steendam, would throw down his pen and run to fetch her.

Next to the director-general, the midwife held the most important position in the colony and everyone knew it, despite the fact that the job had the lowest pay of any official post. (Trijn Jonas's comparatively paltry salary was offset by an exemption from taxation on beer, a free supply of firewood, and the tip she could expect to receive each time she deposited a newly swaddled infant in a father's eager arms.) Midwives constituted the foremost female professional group in Dutch society. In Holland each midwife held a membership in the guild of surgeons and had to follow municipal ordinances and regulations to ensure a high standard of care. To gain their credentials, these women studied advanced lessons not offered to women in any other field. "Midwives are entitled to education in the anatomy hall of the Surgeons Guild," stated an ordinance published by the Delft aldermen in 1656. "When a female body is present the attending anatomist will teach them about the uterus and other connected organs." At the guild, a midwife in training worked with an obstetrics-oriented surgeon called an *accoucheur,* whose skills she could call upon in the event of complications. She carried a modest technical tool kit: scissors, catheter, enema syringe, and materials to tie the umbilical cord, as well as a spice box containing herbs and emollients.

Her training exposed Trijn to key scientific precepts of the day. Like other members of the European medical establishment, midwives learned that the condition of breeding represented a temporary imbalance between sickness and health, proved by its symptoms of nausea and weakness caused by an excess of humors. The planets, theorized the best thinkers of Europe, controlled the growth of the fetus. Still, the child was of the earth, and the baby's nourishment required feeding in the womb upon healthy menstrual blood. The mother must remain healthy in order to birth a healthy child. Thus the first recommendation Trijn offered her patients was dietary: The mother must consume simple food—plain cooked meat, wheat bread, ripe vegetables, and stewed fruit—and avoid any overhot mustard, pepper, or garlic.

Hearsay and old wives' tales also circulated: It was said that drinking sage ale after waking strengthened the female constitution. A woman's congress with her husband must be avoided, most certainly in the first months, so as not to disrupt the growth of the fetus. A woman must religiously wear a piece of sea coral on her person to hold the child fast in her womb. And if Margaret regularly smoothed the birth place with the grease of the capon, she would find the rigors of her labor dramatically eased.

Probably the most important dictum advised that a mother-to-be should avoid looking on "terrible things" to guard against spawning a monstrous child, such as the headless infant conceived by an Englishwoman in Boston. According to the Puritan elders—not representatives of the Dutch Reformed Church, but Protestant authorities nonetheless—that hideous birth came as a sign of God's condemnation of the woman for following Anne Hutchinson, the spiritual outcast (and skilled midwife, as it happened) whose dangerous gospel had led to her banishment from Boston. Then, in Rhode Island, Hutchinson herself gave birth to a barely formed child, said to resemble a fleshy cluster of grapes with lumps for features and lumps for limbs, like a creature that crawled the bottom of the sea. After an investigation, the churchmen declared an exact numerical correlation between the number of lumps—twenty-seven—and the number of heresies for which Hutchinson had been convicted.

The Dutch of Manhattan lived close enough to the Indians to observe how their native counterparts birthed their babies. The Europeans sometimes were amazed by what they saw. The pregnant *wilden,* wrote Van der Donck, "depart alone to a secluded place near a brook [. . .] where they can be protected from the winds, and prepare a shelter for themselves [. . .] where [. . .] they await their delivery without the company or aid of any person." Afterward, a woman would plunge the baby into the cold water for a bath that would not only cleanse its skin but toughen its spirit. When the mother returned to her family, she bound the baby in a snug hide and returned immediately to her usual labors. And although native women labored with no midwife, no hearth fires or fresh childbed linens, these women rarely died, nor did their infants.

By contrast, historians estimate that in seventeenth-century Holland, an astounding 50 percent of all infants never reached the age of twenty-five. In America, too, parents buried a shocking number of offspring, although available evidence suggests the infant mortality rate was lower in the New World. A scholar surveying New Netherland residents of Tappan, New Jersey, between 1688 and 1743 calculated an infancy death rate of 30 percent per family. Female settlers, however, seemed prone to greater fertility than they did in Europe. Colonists surmised there was something in the waters or the air of America that magnified their ability to conceive. It was a commonplace that women unable to have children before they sailed to New Netherland became what one colonial visitor termed "joyful mothers on this new soil."

*W*hen Margaret entered her labor in May 1660, her house was spotless. The plank floors had been swept clear of every speck of dirt, and the tables, chairs, carved benches, and solid wooden wardrobe (known to the Dutch as a *kas*) had been polished. Margaret took part in this cleaning ritual, although the hiring of a servant girl was also customary to get everything in order. Her helper's tasks would include replacing the everyday bedsheets of the mother-to-be with fresh linen imported from Holland. The cache of childbed linen typically included new shifts, caps, stays, and gloves for the mother, and a collection of what one expectant mother noted in a 1698 childbed inventory as "pure fine" infant clothes. A distinct formality accompanied preparations for childbirth, offsetting the earthiness of what would soon follow.

Margaret awaited her labor in the *groot kamer,* the great room, the single room of the house aside from the tiny kitchen wing. In the European tradition, that one space served all the couple's basic needs, including socializing, sleeping, and making love, with a coziness and simplicity that compensated for the obvious lack of personal space. The room would now be employed as a birthing chamber. The May sun still was high, but there was no need to close the shutters, no worries about privacy. No passerby could spy into casement windows set so high up in the wall. Besides, child-birth was a semipublic occasion for Dutch women. Margaret would be attended not only by her midwife but also by friends and neighbors eager to make themselves useful. Modesty typically took a backseat in this group effort, with the exception of the birth mother, who would wear the usual tent of skirts layered one atop the other throughout her ordeal.

Margaret's mother would not be with her. Thus far, the only other Hardenbroeck to reach New Netherland was Margaret's brother Abel, and his presence would have to have been anything but supportive. His passage from Amsterdam in 1659 had been funded by one Conrad van Eyck, who committed to taking on the young man as a shoemaking apprentice on the rather stringent terms that were customary for "an honest shoe hand to perform." From five in the morning until nine at night, Margaret's brother would labor in the "boot-shop, in the tannery and on the tubs," completing a minimum of ten pairs of shoes every week, in exchange for free room, board, "laundry and starching" at his master's home. In May 1660 he was fined forty guilders for breaking windows and making noise in the street. Van Eyck released Abel from his three-year indenture, saying that he and another apprentice "do not work for him, but sit and drink whole nights and then [come] home." Abel was not going to be of much help to his industrious big sister.

As for Maria Hardenbroeck, Margaret's mother, her journey from Amsterdam to Manhattan would not take place for another six months—she would embark on the ship *De Trouw* with her husband on December 23, 1660, when Margaret and Pieter Rudolphus had been married just over a year, with a probable arrival of late February at the earliest, given the typical good-weather eight-week ocean transit plus winter conditions. On that voyage, the couple's twenty-five-year-old son Andries would accompany them, and their new, successful son-in-law would even pay for the young man's passage (the hope was that the Hardenbroecks had only one black sheep son, and that Andries would acquit himself well in the New World). With the advantageous matrimony of their only daughter to a well-to-do merchant, Adolph and Maria could make this late-in-life relocation with a degree of comfort and security that was unusual in the colony. And Margaret's mother would be at her side for future births. But for now, Maria would not be present to perform the tasks that commonly fell to mothers of women in childbirth, those involving technical know-how but also the basic chores of washing and wringing and keeping the fire stoked.

The conjugal *bedde* rode high above the scrubbed wide-plank floor. Its grass-filled under-mattress supported a layer of eiderdown and then the coarse weight of wool blankets. The hired girl had taken the bed pillows outside and shaken them to soften the spot where Margaret would rest her head.

The childbed sheets were, as custom dictated, the finest Margaret could afford. They were coarse, heavy panels woven of flax and hemp, sewn together with a raised central seam. Only the very wealthiest women could afford bedsheets woven on a loom sufficiently wide that they didn't require a seam up the middle; pure linen fabric woven only of flax was more elegant, but the rougher flax and hemp could take a beating and still offer comfort.

Margaret's sweet-smelling sheets had been washed and hung in the sun, then smoothed into place by midwife Trijn Jonas, who bent inside the crimson hangings that cloaked three of the *bedde*'s sides, checking again that everything stood ready. Aromatic smoke rose from a brass plate on the hearth bench, where the midwife burned a healing herbal mixture. Across the chamber, a collection of heavy jugs filled with springwater stood beside the rear wall that Margaret had whitewashed only days before. Keeping the walls a brilliant white was a hallmark of Dutch womanhood. Maintaining a scrupulously clean home was considered a moral imperative, even in the ninth month.

The fire basket waited beside the hearth. This classic Dutch nursery accessory—a bentwood frame surrounding a brazier of wood coals—would serve as a warming rack for diapers once they were washed and spread to dry. A wicker cradle topped with an embroidered satin quilt sat alongside. Tiny clothes had been stacked neatly at the foot of the bed.

The midwife's chair, a central implement of childbearing, already had been carried into the room. The chair's broad back and sturdy arms framing an open seat provided what the women of Holland considered the ideal position for labor. Margaret, like most women, would not stay put in the first hours of labor. Instead, she paced the *groot kamer,* trudging around and around the damask-hung bed that jutted into the center of the room.

By the time Margaret's contractions grew close together, the room had long since grown dark. The rest of the town slumbered, the only noise the faint slap of waves against the wood siding of the canal, the Heerengracht. At midnight came the rattle of the night watchman.

The time had arrived for Margaret's husband to go out into the streets and collect the women who would be her allies. Pieter's task was really the only one for a father-to-be, until the time came for him to lay coins in the hand of the midwife. As Margaret waited for the support troops to arrive, Trijn offered beef broth and made her take some cinnamon water, one of the preferred palliatives for the pains of birth. Then a half-dozen friends arrived. They set out basins, made bawdy jokes to cheer her, and supported her as she paced the room. They poured cool water from the jugs and wiped the sweat from her neck. When the contractions came closer together, Margaret finally settled herself upright into the midwife's chair. The infant, a girl, slipped into Trijn's strong hands.

By the time the night ended, only the midwife remained on hand to debate the repercussions of the drowsy newborn's cut umbilical cord. In a time when people looked for signs in natural occurrences—the depth of rainfall or star alignment, for example—a physical phenomenon as resonant as childbirth was a rich source for prediction and prognostication. Dutch women always had the same discussion at the birth of a baby girl. An infant's umbilical cord, it seems, was better than tea leaves in predicting her ability, when grown, to sate her husband sexually.

*F*or the customary nine days of her *kraambed,* or confinement, Margaret remained at home. She did not work. The fire roared on the hearth despite the warm May weather. Insisting that a piping hot atmosphere improved the healing process, the midwife told Margaret to stay cocooned in piled-high blankets. She dined on strengthening dishes of beef and eggs those first days, along with mugs of strong ale, as her female neighbors doted on her. Trijn Jonas prescribed a tincture of the fresh root of lady's mantle—the "wound herb"—to ease her symptoms after the birth. The baby was lifted periodically from its wicker cradle and placed in Margaret's arms to nurse. After days of rest, men were admitted to the chamber. Some of her male neighbors made their way in and stood quietly by the bed, holding tall castor hats in their hands.

As soon as any new mother's strength had returned, the party known as the *kindermaal* commenced. This was the official celebration of the new mother, Dutch-style, and also her responsibility to host. A feathered cap of quilted satin adorned the head of her husband, the *kraamheer* (literally, birth-man), in the grand tradition of the fashion-conscious Netherlands. (Custom also afforded the new father other privileges during this time, including exemption from creditors, constables, and court messengers, a practice meant to ensure that the mind of his wife, the *kraamvrouw,* remained maximally at ease.) On the front door the new parents hung a lace-embellished birth announcement that resembled an extravagant pincushion, its white ribbons denoting the arrival of a female child. For the celebration, Margaret had dusted the majolica earthenware and taken down the platter on which painted pomegranates danced with clusters of purple grapes, both emblems of fertility and abundance. She spread the *bedde* with her best woven coverlet, with the expectation that guests would sit there and on the few chairs in the room. Margaret covered the table with a deep-pile Persian rug, an item too fine in the Dutch view to lay upon the floor, and set out as many good pressed napkins as she could summon up. Then she loaded the table with food.

The ceremonial centerpiece of the *kindermaal* festivities was the engraved silver caudle bowl and its potent brew. To create the caudle, the hosts simmered equal parts water and Madeira wine with pounds of sugar, oats, raisins, and spices and then left the mixture to steep for hours. The concoction was served hot. Guests would ladle the beverage into their cups using the crook-handled spoons that dangled from the rim of the caudle bowl, and wash down crumbly rusks and the greasy, crunchy *olie-koecken,* which resembled doughnuts studded with currants. The revelers regularly supplemented the stack of *izer*-cookies by the fire, taking turns pressing the *izer* dough into circular irons and holding the long-hinged handles over the hot ashes until the waffles got crisp. The eating, drinking, and drunken talk seemed to go on forever.

Finally at the end of the day Margaret was left alone with Pieter and their baby girl, Maria—named for Margaret's mother, as was the Dutch practice. A threesome, content.

---

*T*he domestic tranquillity would not last long. A scant twelve months after the birth of the baby, Pieter lay dying. At the age of fifty-eight, he was one of the more mature colonists on Manhattan. Not exactly elderly, perhaps, but Pieter was of an age when any number of little-understood diseases—"pestilences" in the jargon of the time—could ruthlessly dispatch a man. While the cause of Pieter's death was not documented, his recorded presence in a courtroom just months earlier suggests his demise was quick. Few accounts detail the specific plagues Dutch colonists experienced (unlike the settlers of New England, who inked heartrending descriptions of their physical trials), but the official records of New Netherland allude to waves of disease that regularly swept the populace.

"Hot sickness" descended upon New Amsterdam in 1658 and again in 1661, the year Pieter died, with an epidemic so severe that the director-general eventually prescribed a requisite day of thanksgiving, fasting, and prayer so the community could assemble to do penance and ask God's forgiveness. (The year 1661 also brought a harvest-spoiling one-two punch of drenching rains followed by searing drought, and colonists prayed for respite from all three.) These scourges followed others: In 1649 Stuyvesant had called for a day of fasting and prayer to combat sickness, probably the "chin cough," or whooping cough. In summer 1655 a plague hit at the height of Indian rampages through the city.

The hot sickness of 1661 was a possible cause of Pieter's death, given that the springtime of that year found him alive and well one month and deceased the next, a typical pattern for untreated illness. Whatever the cause, Pieter's death made Margaret a young and financially secure widow.

Mourning in New Amsterdam, like childbirth, was attended by certain rituals, not all of them somber. As soon as the church bells tolled to announce a passage, aged gravedigger Claus van Elslandt tottered through town, black crepe streamers fluttering from his black cocked hat, informing invited mourners as to the time and location of the funeral. The job of *aansprecker,* funeral-inviter, was one imported from the Netherlands, derived from the Roman tradition. Van Elslandt commissioned the coffin, "cried" the funeral throughout the town, and then walked respectfully before the corpse as the procession made its way to the burial place.

Pieter would have selected and paid for his coffin of fine seasoned cedar long before he lay on his deathbed, as was customary. The coffin's wood was draped in a pall of fringed black cloth. After the dominie came to the house and spoke, twelve men carried the coffin to the graveyard at the foot of the Broad Way, close to the fort. A small procession followed behind. Some men and women had spent the night in the family's *groot kamer,* renamed for this purpose the *doed kamer* (literally the dead-room), to watch over the body. Women joined the men at the funeral service, but the procession to the burial ground had been men only. Later the mourners returned to the widow's house to feast on headcheese, wheat rolls, and thick-cut *doed-koecks,* or dead cakes, flavored with caraway and marked with the initials of the deceased. They drank cold wine and hippocras, the mulled wine whose recipe dated from Roman times.

The funeral feast was expected to be as lavish as the family could afford. Even people who did not have a great deal put aside supplies that would satisfy their guests, often well in advance of their demise. For Pieter's feast, Margaret followed his wishes. She set out a dozen new Gouda pipes in a crock, next to a brass box with Chesapeake tobacco from Pieter's warehouse. Many mourners visited that afternoon, including business acquaintances of her husband whom she had not met before. There had been much distraction around town at the time of Pieter's death. Two women had been convicted of theft—one to be scourged until she bled and the other tortured on a rack. Still, the acquaintances of Margaret's late husband managed to focus attention on his funeral, and her home bulged with guests.

Gowned in dull black bombazine, Margaret distributed an ivory linen funeral scarf to each guest as a souvenir. She wanted to present a fine burial. Yet she would not invest in rings engraved with the date of death or those fashionable silver monkey spoons— round, shallow bowls, handles carved with a simian face—that some burghers gave as funerary favors. In all likelihood she could afford the expense of such gifts, but a brief glance at her husband's accounts revealed that fiscal caution might be a better course. Pieter had not kept his business records as well as he should have and she still had no clear idea what his estate was worth.

Most of the men had come attired in their best, with a small silver-hilted sword at the waist. The women wore soft lace collars and white aprons over luxurious silk and satin petticoats. Those that had them slipped on gold rings studded with garnets, pearls, or diamonds. Most wore white caps that covered the ears, with *oorysaers*, the small gold or silver chains that hung around the back of the head, fastened to cloth on each side with finely worked pearl clips. The more observant clasped Bibles bound with black calf, corners tipped with silver.

The feasting continued for hours. Guests brought tankards of their own, as was usual, though Margaret had offered pewter mugs just in case. As people got drunker, the drinking games grew rougher. Some of the tankards had silver pegs fitted inside that showed the depth to which the drinker should gulp before passing on the cup. Amid the funereal frenzy, baby Maria cried in the smoke.

A brace of barrels sat by the dock, chunky eighty-four-gallon hogsheads, each with a set of initials—*M.H.*—scorched into the wood disk top with a red-hot nail. Following Dutch custom, Margaret had kept her initials and her name after she married, unlike the matrons of France and England who stayed buried beneath their husbands' patronymics. With the death of Pieter, she had been handed a business to run, a fleet to sail, and a good use to put her good name to.

In the summer and fall after Pieter died, Margaret lived with her daughter in the house on the canal. On warm nights she sat out on the stoop as the baby pushed here and there in the street, taking awkward steps in her loopwalker, basically a cage of bent saplings set atop four casters. Soon Maria would be a toddler. Then the child would manage on her own, with only the leading ribbons—streamers that dangled down a toddler's back to be yanked if necessary by an observant parent—to keep her safe.

Margaret's parents were now on hand to help her care for the baby. There was some talk about the two peripatetic immigrants moving on again, perhaps to the recently founded village of Bergen across the Hudson in New Jersey, where Pieter had bought some building lots. The senior Hardenbroecks might be curious about life across the river, but they would not easily leave their daughter while she was recovering from Pieter's death and dealing with the details of his estate. Dutch inheritance laws stipulated that in a case of intestacy, the spouse and children of the deceased must split the estate. As was customary and legal, Margaret would continue to manage Maria's share along with her own until she was no longer a minor.

There was quite a bit to manage. The she-merchant's holdings included not only the Bergen land but also four substantial Manhattan properties. Margaret took over, too, the active import-export business she had built as a partnership with her husband. The trade was lucrative, heavy on tobacco, and she probably could have sold her interest for cash, taken on a partner, or kept the operation at its current level—a good business thumping along at a manageable canter rather than an outright gallop.

But it was not Margaret's desire to cash out or partner up. Nor was she inclined to keep the business exactly as it was. Adding to the one ship Pieter owned, Margaret purchased two more to expand her fleet. She had to sell some of her real estate to make the buy, but she was perfectly within her rights to do so.

Margaret still worked for Wouter Valck, still performed those more rudimentary transactions, as she would until 1664. She had branched out, too, and now executed deals for another merchant named Daniel des Messieres. But her primary work was the business she had undertaken with her husband. Now Margaret alone would uphold the family burgherright, the privilege that had accrued to her in deed if not in name even while her husband was alive. When it came down to it, the title was just a word. More important was her ownership of the fleet of wooden sailing vessels that entered the harbor stocked with casks of oil, pins, wine, or prunes— the ultimate status symbols in this commerce-driven society.

Ships were impressive and their cargo just as important. Without an experienced husband to guide her, Margaret alone would commission the contents of the casks bearing her initials and negotiate for the best deal possible on goods. In just two years Margaret had grown from ingénue-ship-factor newcomer to ship-savvy trading partner to independent wholesale merchant.

To make sure that the casks themselves were sound, she studied the science of tweenery, cask building, the all-important basis of the shipping industry. Some goods could be stacked and bound, and some traveled in trunks. But everything else went by cask. Creating wooden containers that were watertight, airtight, and of a uniform size required master craftsmanship.

Of all the shops along the Strand, the cooper's was the most critically important for shipping products abroad. Margaret learned the fine details, that white pine made the best staves, seconded by cedar and oak. The wood had to be carved properly on the shaving horse, wide in the middle and narrow at the ends, and then heated and trussed with a windlass. The bunghole on the top lid and spyhole bored through the side gave the merchant some satisfaction that he got what he ordered. The craftsman known as the tight cooper created containers for liquids, using staves of white oak and ash, while the slack cooper used oak, ash, maple, hickory, or chestnut to make barrels for sugar, flour, or biscuit. Hickory made the best hoops in this age before the use of iron bands.

In New Netherland, as in Holland, strict laws mandated the size and the quality of all containers. But a watchful eye could still catch errors, and Margaret's eye was nothing if not watchful. Margaret came to understand the exact capacity of each hogshead. A barrel must contain no less and no more than eighty-four gallons straight up. She knew the proper container for a shipment of tobacco: hogsheads of seasoned lumber, "Virginia Size," with six flat hoops around. She knew what to expect, what to demand, of the goods merchants sold her in casks. The contents must be high quality: the pork or beef sweet and wholesome, not "blown," or afflicted with maggots; the flour packed full and level; and the fish of one variety only.

Skins, she learned, also were packed in casks, the beaver pelts cut round to fit the barrel's dimensions, with paws, snout, and tail trimmed off, stacked fifty in a rawhide-bound bale like silky, golden fur pancakes. Each pelt was worth eight guilders in wampum beads. Margaret had watched the packs of furs being off-loaded from up-river ships into the warehouses of the bigger merchants. She had not yet saved the cash to invest heavily in the animal trade. But soon she would.

Margaret still wore black in deference to her late husband; she had for nearly a year. That was the extent of her outward mourning. She had obligations; ships to requisition, inspect, and remodel. Margaret was not the type to let the business wait while she wept at the graveyard.

5

Education of
a She-Merchant

As a child in Amsterdam, Margaret spent her days in a typical schoolroom—a dim space with a low ceiling and wooden benches that lined the floor like church pews, row upon row. The room's four plaster walls were bare. No bright decorations. No alphabet. No numbers on display. None of the colorful maps a proud burgher might hang to enhance the decor of a comfortable middle-class home. Maps were luxuries not allotted for in public school budgets. And that was odd, because the world's greatest seaborne empire had ample reason to educate its children about Holland's expanding role around the globe.

Odder still was the fact that Margaret had a seat at all. The inclusion of girls distinguished the schools of the United Provinces and, it could be argued, propelled the country toward its Golden Age prosperity by equipping a corps of literate, savvy female colonists to build their fortunes in the New World. Margaret and the other girls at school had been awarded the chance—as a right rather than a luxury—to learn the same fundamental skills as their brothers, the reading and writing and arithmetic they would need to make successful lives alongside the men of the Dutch Republic. No higher education was available for girls or women, nor would it be for hundreds of years. Yet the academic opportunity the United Provinces offered its female citizens was extraordinary for its time: Holland was the sole nation in seventeenth-century Europe to offer girls primary education as a matter of course.

Petticoats would not brush up against breeches on the benches of the Dutch schoolhouse, though. Academic instruction was kept as segregated as possible, within the limitations of a one-room facility (modesty notwithstanding the common toilet in one corner). If a school district could afford two salaries, it assigned its female students an instructor of their own. Schoolmistresses tended to have less education than schoolmasters, and sometimes relied upon a program of culturally sanctioned needlework to bulk up the curriculum. Nonetheless, even less intellectual female teachers offered their students a model of classroom leadership. Sometimes two teachers held sway within a single schoolroom, or girls and boys might be seated in adjoining rooms, the advanced students divided from the more remedial, or rich from poor.

Despite the physical cordoning off of Margaret and her peers, the core curriculum in Holland's primary schools remained essentially the same for the sexes. James Howell, an English visitor to Amsterdam, described the autonomy that resulted from the superior education Dutch girls received. Women were expert "in all sorts of languages," he wrote in a 1622 letter. "In Holland, the wives are so well versed in Bargaining, Cyphering, & Writing, that in the Absence of their Husbands in long sea voyages they beat the Trade at home & their Words will pass in equal Credit."

The Dutch Reformed Church established the country's first public schools in the late 1500s. Eventually local governments became collaborators, funding schools and controlling their quality.

At the time of Margaret's birth in 1637, public schools flourished throughout the country, open to all, though attendance was not mandatory. Education had become such a priority that many city governments paid for free instruction of the poor, a clear rejection of the idea that some citizens had little need for academics. Attendance was expected every single day of the year, with the exception of Christmas, Easter, the holiday called Pinkster, two weeks during the dog days of summer, and Wednesday and Saturday afternoons. Children went to school even on Sunday mornings, then accompanied their teachers to church, which provided further exposure to the Bible passages they already had memorized. Girls scowled on their unforgiving benches and struggled not to squirm, just like boys, from the time they started school at age seven until their book learning ceased at about the age of twelve.

Children completed elementary school in stages. First they learned reading, then writing, which included instruction on how to wield a fountain pen steadily enough to sign with a respectable cursive rather than with a blotchlike mark. Developing a fine hand in Roman and Gothic scripts had become an essential component of early education—a reflection of the increase in personal letters and messages as the century progressed and of the importance of written communication in business transactions. Basic rhetoric, too, received attention.

Mathematics followed, and it was at this point that some children departed the classroom; parents pulled them out to acquire a trade or work as household servants. Students who remained in school learned practical math skills to do basic bookkeeping or execute an official bill of sale. Learning to read, write, and do figures was primarily vocational training. So was gaining a grasp of foreign tongues. A person in business needed to know at least a smattering of languages to communicate with customers, suppliers, employees, and commercial rivals, though a classroom syllabus in languages only could be pursued after primary school at a Latin School or a French School. Neither was open to women. Of course trade itself offered opportunities to master different tongues. Holland's high level of linguistic proficiency comes across in the comment by one seventeenth-century Portuguese envoy at The Hague: "There is not a cobbler in these parts who does not add French and Latin to his own language."

Its schoolhouses may not have been fancy but the system worked: By the end of the century, Holland had the lowest percentage of illiteracy in Europe. In 1630, 57 percent of Amsterdam's bridegrooms and 32 percent of brides were able to sign their marriage certificates; in 1680 the numbers had risen to 70 percent of men and 44 percent of women. While the percentage for Dutch women was lower than for Dutch men, scholars agree that literacy for women surpassed that of any other European nation.

New Netherland had a comparable focus on schooling, with high literacy rates for women and men alike. The colony's first paid academic instructor arrived as early as 1624. Number one in a string of teachers with varying qualifications, Bastiaen Jansz. Krol was hired not only to convey the three R's in a classroom but to baptize babies, bury the dead, perform marriages, and deliver sermons as needed. His educational offices were available to native as well as colonist boys and girls, an inclusive approach that would continue under Stuyvesant, who would appoint a teacher specifically to instruct the offspring of the colony's enslaved Africans. Manhattan's educational system took a shaky turn with the appointment of a loose cannon named Adam Roelantsen, who moonlighted as a launderer (he ran a *bleeckeryen,* or bleachery, where linens were laid out on grass to whiten under the sun) and spent his nine-year teaching stint suing and being sued for slander.

The emphasis on education produced literate colonists. In mid-seventeenth-century Albany, 66 percent of men and 49 percent of women could sign legal documents with their names rather than a mark. Between 1664 and 1695 the percentage rose still further, with 86 percent of men and 50 percent of women signing their own wills in colonial New York. That the tradition of education for women continued for the descendants of the first Dutch settlers was confirmed by one traveler's description of a school in Albany in 1744 that held "about 200 scholars, boys and girls."

Girls in traditional Dutch classrooms were treated differently in one arena of their educational experience: discipline. Misbehaving boys received a whipping with a strap, or even a turn at the post like their elders. Schoolgirls who transgressed suffered a more humiliating outcome: They were beaten and kicked by their instructors. Or a teacher might command a female student to take her seat on a specially designed cushion, tailored of leather, with inset metal tacks pointing up.

That Margaret survived the punishments meted out to girls does not make her special. Nor does her grasp of reading, writing, and math, a gift the fatherland customarily granted female students. Like so many others, she came to possess the bare minimum required to distinguish herself in the world if she chose to. Each Dutch woman of Holland received an identical set of tools.

What she did with it was up to her.

$\mathcal{M}$argaret spent most of her childhood with her parents and three brothers—Johannes, Andries, and Abel—in Amsterdam, the white-hot commercial center of the Netherlands and of Europe. Her parents were immigrants from Germany, Holland's closest neighbor. Elberfeld, the birthplace of Margaret and her parents before her, was a time-burnished place. First documented in 1161, it received its town charter in 1610 and thrived due to its position on the banks of the Wupper River, where the local yarn-bleaching industry flourished. The Dutch favored white in their textiles, the brighter the better—white ruffs, white cuffs, white petticoats, white caps, and stack upon stack of folded household linens, which must be spanking, sparkling white. Bleached white was clean and clean was a moral imperative.

By the early seventeenth century, Elberfeld had grown into a trade-charged metropolis known for its embrace of foreigners, especially the Dutch, who represented such a robust sector of the community that the Elberfeld Protestant Church kept a separate record of Hollanders baptized or married under its auspices. Transit of Dutch goods through Germany along the Rhine carried word of economic opportunities in Amsterdam, which might entice a family looking to expand its fortunes.

At some point in her childhood, Margaret's parents moved the family to Amsterdam, a choice that could not have been easy and showcased an uncommon determination and ambition. Margaret's parents were not refugees, forced to flee because of religious or political persecution, though exposure to an epidemic of plague in 1631 surely figured in their relocation. Most important, as traders they had an appetite for the kind of opportunity the metropolis of Amsterdam would give them.

Once her parents decided to become Dutch citizens, Adolph and Maria made another choice, a classic immigrant decision to revise the family's surname. Hoddenbruch, their ancestral German name, became Hardenbroeck, a designation that bore a phonetic resemblance to the original but enjoyed a special resonance in the Netherlands because of its identification with a local family, the noble Dutch clan of von Hardenbroeck. Since the name Hoddenbruch, by contrast, most likely derived from a farm-estate on which Margaret's ancestors labored as common tenants (a typical naming practice in feudal Germany, where Hoddenbruchs still can be found), trading their last name in for the lordly Hardenbroeck gave the family an instant step-up in status, if not fortune.

Making the move from Elberfeld to Golden Age Amsterdam proved instructive for Margaret. The city was, in the phrase of one scholar, "explosively mercantile." As Margaret's adolescence dawned, the metropolis swelled with people of every stripe, and displayed conspicuous wealth in a wraparound backdrop of luxurious clothing, grand houses, and fleets of majestic sailing ships.

As she matured, Margaret also absorbed lessons by studying her mother and the other women of Amsterdam. In the markets and on the quays, her role models were "frugal, busy, and always doing something, not only the housework," as the Italian merchant Lodovico Guicciardini described them. Amsterdam's female citizens bartered decisively, unapologetically, noisily, and often more forcefully than the merchants with whom they did business. The women of seventeenth-century Holland had an unorthodox approach to the opposite sex compared to their European counterparts. Wives, it was commonly known, ruled their households like queens, telling their husbands when and where to remove their shoes, even dictating which rooms men were permitted to enter. An English naturalist reported that it was customary in Holland for married women "even of the better sort" to kiss men—men who were no more than acquaintances—upon greeting them or taking their leave, a far-too-forward gesture as far as he was concerned. Dutch women were, he said, "more fond of and delighted with lasciviousness and obscene talk than either the English or the French." Another English writer marveled at the liberties the women took for granted: They ice-skated through the night, relieved themselves in public, even went tavern-hopping with their friends. One activity, however, seemed to make the strongest impression: The women of Holland engaged openly in conversation with men as they walked with them in the streets of the town, unapologetically, sans chaperone.

These day-to-day activities had their more refined reflection in the ideas put forward by Holland's cutting-edge intellectuals. Prodigy Anna Maria van Schurman explored some of the more provocative ideas of her time in 1639, when she published her famous celebration of the aptitude of the female mind. Her work was translated into English two decades later as *The Learned Maid, or Whether a Maid May Also Be a Scholar.* She had made a splash in the intellectual pond of seventeenth-century Utrecht with her metaphysical musings, copper etchings, and proficiency in Hebrew, Greek, Arabic, Syriac, and Latin. This female wunderkind assumed the care of two blind aunts for most of her life, nursing them while producing her creative canon and still somehow managing to entertain the stream of visiting scholars and aristocrats who came to pay her homage.

Another theorist, the internationally known physician and popular writer Jan van Beverwyck, published a work titled *On the Excellence of the Female Sex.* He stressed the military accomplishments of female warriors in the Netherlands' prolonged war for independence, citing the wife of a small city magistrate who led a legion of female soldiers in 1576, and the folk hero Kenau Hasselaer, a Haarlem matron-commander whose housewife-troops resisted the Spanish siege of the city with cutlery and soup cauldrons. "To those who say that women are fit for the household and no more," wrote Beverwyck, "then I would answer that with us, many women, without forgetting their house, practice trade and commerce and even the arts and learning. Only let women come to the exercise of them and they shall show themselves capable of all things."

Even the Dutch Reformed Church prescribed equality for women. Margaret's family ascribed to its teachings—they attended services both in Elberfeld, where the presence of the Hoddenbruch family has been established in church records, and on Manhattan, where documents reveal the membership of Margaret, her parents, and her brothers. The Dutch church shared some scriptural elements with Puritans and other Protestant denominations, but differed significantly in its teachings regarding men and women, advocating mutual respect rather than wifely submission to a husband and suggesting that companionship, *gemeenschap,* constituted a primary reason for marriage. According to the church, a proper wedded relationship featured the husband as literal head of the household, but he must demonstrate his worthiness for that role by accepting management of the household as his wife's responsibility and treating her with respect. Calvinist doctrine even emphasized the importance of physical affection as the foundation for a godly union, with fornication in abundance to provide the godly mortar (sex should satisfy the female fornicator, too, said the good church fathers).

Church teachings, the intellectual climate, and intense commercial activity—all helped define women's lives in Amsterdam as Margaret came to maturity. In addition, Dutch law offered women a measure of invincibility. Holland's legal system was fairer to women than any other in Europe.

An unmarried woman in Holland, a maiden or widow, could expect to be treated under the law precisely as a man would be. If married, she determined her own rights within that relationship, according to the provisions of the nation's unique civil laws, by selecting one of two different types of marriage.

If a woman chose the *manus* marital option, she accepted the status of a minor under the guardianship of a husband, who would henceforth serve as her legal representative. She had no standing in court. She could not defend herself; her husband must appear on her behalf if she was charged as a defendant. She also could not institute any proceedings against another party or enter into any contract without her husband's authority. (Any payment for such a contract must be made to her husband.) The practice under this marital option—time honored, but distinctly less popular in the seventeenth century than in previous eras—was for a woman to pool her possessions, including debts and future profits and losses, with her husband at marriage as part of a community of property managed by the husband. The family finances would be entirely under his control.

Three features commended life under *manus*. First, the legal custom offered a leg up for a woman who entered a marriage with no material resources of her own. If her spouse perished, the shared community of property would be cleaved in two, and she could rely upon half going to her while the other half went to her husband's heirs. Similarly, *manus* allowed a woman to assume the quite possibly loftier rank of her husband's family, a marker that was worth something in a still-feudal world. Finally, under *manus* a wife's financial liability was limited in the event of a failed business deal on the part of her spouse—with the legal reasoning that if she didn't have the authority to make the transaction, why should she take any responsibility for it? In practice, it was a respectable and frequent course of events for a widow under *manus* to simply lay the house keys on the coffin of a deceased husband who had squandered her share of the community of goods and literally walk away, free and clear from any claims on resources she never had any control over. A wife's sole authority under *manus* was the power to operate as domestic manager, a role the courts honored to the extent that it could not be denied her without judicial action.

Manus was similar to the marital stipulations of English Common Law, the code that influenced the legal system of almost every European nation. England's central practice of coverture placed a woman automatically under the guardianship of her husband and, like *manus,* forbade her from bringing lawsuits, entering into contracts, or making binding decisions about her own property. By the seventeenth century, an elaborate intellectual framework shored up the coverture mandate of secondhand status for wives; at its center was the notion that a proper English family was a microcosmic kingdom with a godlike husband crowning the hierarchy and wife, children, and servants following in declining status. It was only logical that the man should head up the family if his wife, according to Anglican doctrine, was "a weak creature not endured with like strength and constancy of mind." The constrictions of English Common Law did not even pass away when a husband did. When a woman married, her husband automatically owned her personal property, including jewelry and clothing. If he died intestate, she would inherit just one-third his personal estate (minus funeral costs and debt) and the use of one-third his real estate to live on but not to sell (his heirs got it after she died), a pittance legally known as a "dower," more familiar as a dowry.

Holland provided an alternative for women who disliked the *manus* option. Under the precepts of what was termed *usus,* a wife retained all the rights she had as a single woman (the same rights as any Dutch man), and the relationship of husband and wife became a partnership of equals. A wife was free to appear in court as either a plaintiff or defendant, or to represent her husband before the judge. To secure the advantages of a marriage under *usus,* a woman made a prenuptial agreement with her husband that circumvented the community of property and the husband's extraordinary powers over his wife.

The legal conventions surrounding inheritance, too, distinguished Holland as a uniquely egalitarian society. At the death of either spouse, in cases of intestacy, the survivor could expect to receive at least 50 percent of the estate, even under *manus,* regardless of how much property a spouse brought to the marriage, with the remainder going to the heirs of the deceased. (Wealthy widows were commonplace in Holland and the Dutch colonies.) By the same token, Dutch law prohibited parents from relying upon gender or birth order to deed their offspring property in wills. That meant daughters were not arbitrarily deprived of an inheritance. Dutch inheritance strictures had their mirror opposite in England, where firstborn sons received all of a family's major property holdings, such as land and houses. Daughters, in contrast, customarily received only household goods (flatware and furniture). Female heirs often faced the future after a parent's death without a home or the assets to obtain one. Widows customarily came away from high-flying marriages with only the dowry they had brought to the altar.

Dutch law also protected unwed mothers, stipulating that a woman impregnated outside marriage could either prosecute the alleged father in a paternity suit, a *vaderschapsactie,* or force him to marry her. If he was already married, she was entitled to demand a dowry and compensation for childbirth expenses, as well as child support. The mother had a good chance of winning some form of support, report historians of the period. A husband's adultery, willful abandonment, or contraction of a venereal disease gave his wife grounds for requesting a divorce. Spouse abuse also received attention under Dutch law. And if a wife believed her husband was squandering their property, she had legal recourse to request her half of the estate, along with her dowry in full. Regardless of whether she married, and in distinct contrast with English law, any Dutch woman could institute legal proceedings against any individual, female or male, even her husband.

Conservative and pragmatic, confident in their equality under the law, Margaret and many other Dutch women applied their schooling and street smarts to shaping New Netherland. Commerce was in their blood. They also took a stubborn pride in the domestic traditions of the Old World, the female traditions, the ones that would last. Their strength was rooted in home, hearth, and the marketplace, and carried over to America as carefully as a packet of garden seeds.

6

A Marriage of Love and Trade

Like other widows of New Netherland, Margaret wasted little time finding a suitable second spouse. New Amsterdam's Dutch Reformed Church published the banns for her marriage to Frederick Philipse, a carpenter in the employ of the Dutch West India Company, on October 28, 1662, just a year and a half after Pieter's death. Like his new wife, Frederick would go from hungry immigrant to colonial kingpin in the course of a lifetime. And like her, he married extremely well at a crucial moment in the building of his fortune.

We have no record of the couple's earliest meeting. But their activities suggest that Frederick Philipse had been a visible presence in Margaret's world in the years preceding their wedding. Frederick spent many hours in Bergen during the founding of the town to assess the work completed there. He also frequented the Strand, especially the taverns along Pearl Street, where he socialized with skippers and ship carpenters, and the shipyards, where he had outfitted a good-size sloop for the West India Company. He may have started out a woodworker, but like everyone in New Amsterdam he wanted a piece of the action, the business of trade. In 1660 he received permission to charter the ship and sail it down the east coast to take on a load of Virginia tobacco. The resolution that allowed him to take the ship describes Frederick as "late the director's carpenter," a clause that suggests the artisan's special status in the colony. He was already thirty-eight and his knowledge of carpentry gave him an advantage over most wholesalers in making sure any ships he chartered were seaworthy.

Frederick Philipse landed in New Netherland in 1653, six years ahead of Margaret. Billeted under his given name, Vrydrich Flypsen, he had sailed from Bolsward, Friesland, an inland shipping and trading center in northern Holland where he had grown up the middle-class son of a roof slater. (The era had not yet come for standardized spelling of names; archival documents show at least thirty variations for Flypsen, including fflyson, Flypsie, and Vlypzen. Frederick gradually anglicized his name, though he used the initials *VF* throughout his life.)

In an era of sea travel and empire building, men with superior woodworking skills were more than just tradesmen—they were celebrities of the work world. That translated into opportunity for Frederick. He first enters the historical record with a 1653 Company order appointing "Frederick Flipsen" to appraise a house and lot in New Amsterdam. Stuyvesant's campaign to restore the colony's damaged structures led to Frederick's early success. There also was a constant demand for new houses, warehouses, workshops, and market buildings.

Frederick's world was in transition. Clearly the colony still needed his carpentry skills. Yet he had been directing a good deal of effort, if not his all, to a new career, a shift that dovetailed nicely with his engagement to an up-and-coming she-merchant. In the census of 1657, he was listed as "Fredrick Flipsen, Carpenter." His status remained "carpenter" in March 1662. By October of the same year, the month the church announced his marriage banns with Margaret, Frederick's name was listed as the more impressive "merchant."

Before Frederick and Margaret finalized their vows, Margaret chose to establish their partnership according to *usus,* crafting the age-old prenuptial contract that explicitly denied a husband unlimited power over his wife. (Like so many of her New Netherland documents, her marriage contract with Pieter Rudolphus did not survive.) As a she-merchant who already ran an independent trading concern, Margaret needed the control of her finances. Entering into her marriage under *usus* ensured that the property she brought to the marriage—the house lots in Manhattan and Bergen; ships that now included the *New Netherland Indian, Beaver, Pearl,* and *Morning Star;* and her furniture, plate, and linens—would remain hers. She would continue as "a free merchant of New Amsterdam," as one court transcript described her. She would be able to make contracts, to sue and be sued in court, and to trade independently. Marriage would not impose the restrictions under which her sisters in Boston or Virginia were laboring.

Marriage under *usus* was the general rule for Dutch brides in New Netherland, as it was in Holland. Where this couple departed from the norm was in the proportion of wealth the bride brought to the union. Her fortune outweighed his, by far. Was Margaret mainly a meal ticket for a hungry burgher? Perhaps in part. Her own first marriage to an older, established merchant probably had contained some degree of calculation. Frederick could certainly use her wealth and status as a base for his increasingly lofty ambitions. His own estate included mainly his artisan's earnings and whatever yield he had begun to see in the tobacco trade. On balance, his ledger sheet might even reflect more debt than profit at this transitional phase of his life.

But gold digging could not have been the whole story as Frederick and Margaret came together in the fall and winter of 1662. First of all, it is hard to believe that canny Margaret would tolerate a bounder. Second, Dutch tradition took seriously the ideal of marriage between two individuals who genuinely respected and felt attraction for each other. Frederick's attraction to her might show some real character on his part; history suggests that on an interpersonal level, nothing about the woman was easy.

The feeling surely went the other way as well. Margaret must have appreciated his creativity and the skill he brought to his work, his sterling reputation among his peers, and the impressive contacts he had forged throughout the province. Everyone in the colony knew Frederick Philipse. He was the governor's go-to man in the building trade, a standout in the colonial hierarchy even without the formal imprimatur of the great burgherright. On top of all that, Frederick's pursuit of the coastal tobacco trade evinced initiative. His fortune paled beside hers, yes, but he showed the potential for greatness. Perhaps Margaret's shrewd nature let her fall in love with what might be: A spouse could do a lot to shape her mate's potential.

One more circumstance of the Philipse-Hardenbroeck nuptials underscores the probability that the marriage rested on a foundation of genuine feeling. When the couple came together, Margaret's daughter Maria was two-and-a-half years old.

Shortly before they wed, Frederick promised that "out of special love and affection" he would "keep said child as his own," and make her an heir on the same par as any children he might have in the future. Frederick made those declarations in response to an inquiry by Orphanmasters, a colonial institution that demanded an accounting of how Maria would be provided for after her mother's remarriage.

The Orphanmasters was a concept imported from Holland, where municipal care of unfortunates constituted a venerable tradition. Golden Age painters regularly received commissions to portray the advocates who were charged with oversight of charitable institutions. Each portrait shows a passel of society's leading lights in their coal-black finery, the men in beaver hats and beard tufts, the women with caul-like caps and tight-lipped expressions, all shrouded in shadow with only their white faces emerging from the gloom. The great Calvinist goodness of these regents was typically underscored by the presence in one corner of the canvas of a grateful child either wearing the distinctive uniform of the orphan asylum or with its hands bound up in a leper's mittens. Given the colony's chronic fiscal problems, which made even the upkeep of Fort Amsterdam a struggle, creating a needy children's institution was not first on the Company's agenda. But when New Amsterdam received its charter in 1653, the new municipal government appointed two men to "keep their eyes open and look as Orphanmasters after widows and orphans." Yes, Maria still had a living parent, but orphanhood had a rather wide definition in New Amsterdam.

And so it happened that the Orphanmasters' docket in the fall of 1662 included the question of provisions the late Pieter Rudolphus de Vries had left for his only offspring. He had no will, so Dutch law provided for an even split of his estate between his wife and infant daughter. But what did that mean exactly? Clarity was required, and time was of the essence. When Orphanmasters Martin Kregier and Cornelis Steenwyck fastened upon the case of Maria Rudolphus de Vries, the banns of the Philipse-Hardenbroeck marriage already had been published. On November 23, the Orphanmasters requested that Margaret render an accounting of her late husband's estate, specifically its provisions for their daughter, demanding a "statement and inventory of the property, left by said Pieter Rudolfus and to be settled on his child as inheritance." She had eight days, they said, to provide the information.

A week and a half later Margaret appeared again before the Orphanmasters. She said she could not deliver a precise account of

Pieter's debts or credits because his affairs in Amsterdam had not yet been resolved. She was awaiting paperwork, expected to arrive any day on a boat from Holland. (It had already been twenty months since Pieter's death.)

The case slid into a deadlock. The Orphanmasters would not retreat from their insistence on seeing Pieter's accounts before the wedding was solemnized. Finally, on December 18, Frederick put his signature on a legal agreement that the Orphanmasters said they would accept as a viable substitute for Pieter's estate. With this contract, Frederick pledged himself legally and financially to protect Maria's well-being so that his union with her mother could go forward.

This was not a small thing, Frederick's embrace of the child as a joint and equal heir. And afterward he took the further step of legally adopting the little girl. As was his right, and in the ritual that was by now customary for a family whose members took pains to conform to their surroundings rather than hold themselves apart— Hoddenbruch to Hardenbroeck, Flypsen to Flipsen to Philipse— he changed the baby's name. Henceforth, Maria Rudolphus de Vries would be known as Eva Philipse. With the financial agreement, the legal adoption, and the name change he made her his own. Why? It is possible that adopting the child was an ultimatum Margaret laid out for all her potential suitors. Still, it seems from her repeated performances before the Orphanmasters that she was actually trying to resolve Pieter's estate issues, and therefore Maria's status. It appears as though Frederick got fed up with waiting to marry Margaret. It is hard to believe that he would go to these pains to adopt this child—even investing his hard-earned money to do it, now and forever—unless he had some genuine love for her mother.

Arguing her case before the Orphanmasters constituted only a fraction of the energies Margaret expended in court during the period after Pieter's death and the first years of her marriage to Frederick. Because in fact Pieter's affairs *were* in considerable disarray, probably aggravated by his sudden death. Since laying her first spouse in the grave, Margaret had been embroiled in sorting through his debts and assets. Resolving Maria-Eva's inheritance was one thread in a tangled skein of financial anomalies—anomalies that Margaret frequently worked to her commercial advantage. The Orphanmasters episode was no exception. In the end she never had to open her books to the authorities and so avoided any official oversight concerning how she administered her daughter's property or her own.

Sometimes it seemed Margaret spent more time in court than the magistrates themselves, collecting on her first husband's debts and defending his estate. In Margaret's time, the city's court of law was comprised of the *schout,* who wore the hats of both sheriff and prosecutor, as well as two comayors called burgomasters and five aldermen. The panel served as judge and jury. She spent more than three years in and out of court battling a plaintiff who tried to seize all "goods, actions and credits belonging to the late Pieter Rudolphus."

Court records show Margaret appeared frequently to defend her deceased husband against charges of nonpayment, but also as a defendant in cases involving her own business practices. In 1662 Johannes de Witt alleged that Margaret had not paid the fifty-one guilders and ten stivers she owed for wampum; apparently the panel looked on her explanation favorably because there is no record that the court ordered recompense. In January 1664 an Amsterdam notary recorded the intention of "Margrieta Hardenbroeck, from Evervelt [Elberfeld], wife of Frederick Philipse, free merchant on New Amsterdam in New Netherland," to pay a debt to local widow Trjntje Willems, "tradeswoman in Amsterdam," for linens she was exporting to Manhattan for sale. Margaret swore that she would deliver the amount of 529 florins to Willems within eight months of signing.

In New Netherland, such transactions comprised a central part of doing business, for blacksmiths, bricklayers, tapsters, and farmers as much as traders. A business owner's skill in arguing a case in court was key to commercial success. Margaret cut a persuasive figure in the courtroom. She did not win every one of her cases, but she was a reliable presence on the public scene, a person who could be counted on to defend what was hers.

New Netherland had a number of power couples that grew their fortunes as partners, among them Jeremias van Rensselaer and Maria van Cortlandt (one of Margaret's trading partners), and Schuyler and Alida Livingston. Merchant Cornelia Lubbetse, credited with importing the first cargo of salt to the colony, married Johannes DePeyster in 1651; she died at age ninety, leaving seven properties on Manhattan. But no colonial duo matched Margaret Hardenbroeck and Frederick Philipse for mutual involvement in a large-scale commercial enterprise. Their professional dependence began at the inception of their union. When it suited Margaret to have separate representation in court—she might be conducting business elsewhere, leaning on a delinquent buyer, say—she assigned her husband power of attorney. Frederick returned the favor when the reverse made sense.

Like his wife, Frederick was dogged, often single-minded. On one occasion, he and three associates set out to trade on the colony's outermost frontier, about one hundred miles north of Manhattan on the bank of the Hudson River. It was December 1663, the very end of harvest season. After taking care of business at the village of Wiltwyck, the men drove six wagons full of purchased grain out the settlement's gated entrance and down to a protected harbor area to ship their product to New Amsterdam for resale. All standard practice. But Frederick and his companions did not wait for a military convoy to escort them through the surrounding woodlands as required by law at that time. Signs announcing the regulation—designed to protect colonists from ambush by the Esopus Indians—were posted all over the place. They were hard to miss.

The regulation emerged from a series of events that made the preceding year one of the most extraordinary in the history of the colony. Settlers had witnessed a terrifying meteor come "out of the bosom of the moon" to cross the sky, then experienced as hard a freeze as any in memory (horses crossed the rock-solid Hudson for months). Spring brought torrential rains and flooded fields, followed by an uncontainable smallpox outbreak. In June the Esopus renewed attacks on colonists after a three-year truce, culminating in a massacre in Wiltwyck that left, in the words of eyewitness minister Hermanus Blom, "dead bodies of men [. . .] here and there like dung heaps on the field and the burnt and roasted corpses like sheaves behind the mower." More than twenty colonists died, twenty-five went missing, and forty-five were taken hostage by the Indians. The village lay in ashes. In retaliation, Dutch soldiers with the aid of a healthy complement of mercenaries annihilated the Esopus, with the exception of a few who escaped. And settlers began to rebuild.

Frederick's wheat wagon escapade took place several months later. To give the four traders credit, they had managed to sniff out what little stores of grain survived the inferno of Wiltwyck, and doubtless struck an excellent deal. But why would Frederick flaunt martial law to move the wheat when rational minds might consider protection a welcome thing? He and his cronies had acted "in disobedience to the orders of the Director-General and Council," as the official complaint about their act phrased it, even with "great danger imminent." The answer, according to the recorded statement of the accused, was that the men simply found it inconvenient to wait. They knew the military escort was there, at Wiltwyck's (well-reinforced) gate, but the wagons had already started down the road, and the soldiers were making their way lackadaisically behind them. Frederick, it appears, just could not stand to take any more time to get his merchandise to market, even when caution might mean self-preservation. He probably weighed his chances and figured the director-general's men had cleansed the forest of most adversaries, if not all. So why worry? The twenty-five guilder fine charged to each scofflaw in his group amounted to far less than the profit the shipment would bring.

Margaret had met her match.

*W*hile Frederick was off inciting the wheat insurrection at Wiltwyck, Margaret was approaching the end of her second pregnancy. By March 1664 she had a healthy baby boy, a child the couple christened Philip. Eva now had a half brother four years her junior. Margaret was busier than ever, caring for her two young children, along with appearing in court, buying and selling to equip her fleet, and periodically taking off for distant shores for months at a time. Even with a husband-partner in trade, juggling it all was surely a challenge.

It must have been a relief that nearly all of Margaret's immediate kin were in Manhattan. Her mother and father had remained in the New World and joined the Dutch Reformed Church at New Amsterdam in October 1662. The stretch of time between Maria and Adolph Hardenbroeck's initial arrival and their becoming congregants, about a year and a half, suggests a certain amount of indecision. Perhaps the couple had planned the trip as a friendly family visit rather than as a permanent relocation: Colonists did cross the Atlantic more readily than the prevalence of shipwrecks, storms, and pirates might lead a more cautious era to expect, sometimes traveling for trade, sometimes to attend to a death or some other family business, sometimes to retrieve children or a fiancé. Often a person would reach his destination only to circle right around within weeks or months and return to his starting point, be it Amsterdam or Manhattan. Or perhaps they were preoccupied with Margaret's newborn and with their own wayward son, Abel, who had been in one scrape after another since arriving from Amsterdam years earlier as a shoemaker's apprentice. Most recently, Abel had been fined fifty guilders for "force and violence" against his employer and condemned to three months solitary confinement. Following this episode, Abel escaped to Amsterdam and married a young widow, Anneke Meynderts. Together they returned to New Amsterdam and had a son. Margaret's second brother, Andries, also a trader, still lived in the colony. Finally, in January 1664, Johannes, the eldest, followed his younger siblings and parents to America, departing Amsterdam on the ship *De Trouw* with his wife, Urseltje Duytsman, and their four small children.

By then, the senior Hardenbroecks had relocated to Bergen, where they almost immediately took an active role in the development of the area. Adolph received an appointment as magistrate, a role that suggests he was held in some esteem in the community. By 1664 Maria and Adolph were members of Bergen's Reformed Church. Margaret still involved herself in the New Jersey settlement her first husband helped found. Her name appears on a petition of thirty-six voluntary contributors—she is one of only two female members of the roster—who pledged various sums of "guilders in wampum" to bring a clergyman to the community.

Margaret's parents made the move to New Jersey at an age that signaled the start of the golden years for the average colonial settler. Adolph was fifty-five, Maria not much younger. But for these adventurers the move out of Manhattan probably constituted just the tonic they needed. They had a piece of land, courtesy of their successful daughter, and a modicum of security in the little stock-aded village of Bergen. They could easily reach Margaret and their grandchildren on the new ferry service that shuttled passengers across the Hudson River in just about thirty minutes.

Margaret was involved in a relocation of her own. With two children in tow she moved her household from the odd-shaped yard on the canal to a property Frederick owned on the corner of Stone Street and Het Markvelt. The house occupied a handsome, level lot that Stuyvesant and the city council granted Frederick in 1658 to reward his work on behalf of the colony. It was prime real estate, a clear move up, this switch from the Ditch to the Road. It was also a finer residence, taller, two-and-a-half stories rather than a single floor with a storage loft. Wood clapboards, once again, clothed the new abode—not yet the coveted brick—but if its representation on the 1660 Castello Plan is a guide, the structure had more heft than nearly any neighboring domicile. The exception was the Van Cortlandt compound down the block, with its mansion and hulking brewery buildings that sent keg-stacked delivery wagons lumbering out into the street all day long.

In the move to the Road, Margaret gained a larger kitchen, one she could now afford to stock with imported cookware. Despite the increasing plurality of New Netherland in the early 1660s, and the fact that some everyday material goods were being produced on American soil (brick, glass, and ceramics among them), the Dutch ways still held strong. A peek in the swung-open Dutch door of a New Amsterdam kitchen during the first years of Margaret's marriage to Frederick would reveal a kitchen outfitted essentially as it would have been back in Holland.

A massive trove of seventeenth-century ceramics discovered by twentieth-century archaeologists in the silt at the bottom of the East River, thought to be the remains of a shipment from a potting center at Bergen-op-Zoom in Holland, shows New World housewives' predilection for the same kitchen implements their mothers had used for centuries. A carpet of shards of *grapen,* vessels that were used for both cooking and eating, covered the river bottom, where they had been relegated after shattering in transit. Colanders glazed in bright colors were another favorite; a kitchen could not have too many.

Margaret came to her second marriage with a store of domestic goods. On her hearth sat *grapen* galore. As a widow, mother, and relatively mature woman in her twenties, she did not need to start from scratch. Her task on Stone Street would be to approach her household with an eye to organizing and increasing its bounty.

To that end, Margaret would continue to amass her cache of linens, adding to the all-important *kas* that had been bundled ever so carefully in blankets and conveyed by rattling oxcart to the new house. In keeping with the fashion in *kasten,* the front of her freestanding cupboard was painted in the technique known as grisaille, gray wash, which artists had used since the Middle Ages to simulate carved marble. (The use of grisaille work on *kasten* is an approach unique to furniture makers of the bountiful Hudson Valley, judging from examples that survived from the era in Holland and America.) The two door panels of the *kas* boasted elaborate trompe l'oeil pendants of cleft pomegranates, bulbous pears, mounded grapes, and delicate quince suspended with ribbons in parallel make-believe niches—brushstrokes on an oak canvas that art historians judge to be the first still lifes rendered in America. More meaningful for a woman like Margaret, though, was the symbolism of this decoration, which made a potent talisman for fertility and domesticity. The cupboard took its place on wooden spheres the size of cannonballs at the center of her new home, in her *groot kamer,* where it exhibited her taste and social standing, the impeccable organization of her home, and, in sum, her sheer female competence.

Unlike the loom-centered lives of their colonial kin in New England, the women in New Netherland seldom did any spinning and weaving. In fact, authorities explicitly forbade citizens in the

Dutch province from engaging in either craft. The West India Company was intent upon preserving the commercial interests of the United Provinces, and that meant preserving a monopoly on the household linens famously manufactured in the Old World. This policy also supported the commercial interests of wholesale importers such as Margaret, who began to buy and sell vast quantities of Dutch dry goods in the late 1660s. The authorities aimed to stop colonists from trading homespun cloth with the Indians by forcing the settlers to purchase imported textiles from the Company instead. Two other staple chores, baking and brewing, also were largely omitted from women's daily lives because bakeshops and breweries were licensed businesses overseen by the municipal authorities. A Manhattan woman certainly produced some specialty baked goods in her own kitchen, and diarists recorded early recipes for such alcoholic decoctions as raspberry brandy. But a homemaker had only to walk to the corner store for beer and bread, the staples of the seventeenth-century larder. The prohibitions worked in her favor: Margaret could keep her *kas* stocked with an assortment of immaculate, bleached, starched linens to rival any Holland housewife, without frittering away any of her valuable time at a spinning wheel or loom.

As a trader Margaret had connections that increased her access to the highest quality textiles, in all the subtle gradations of white: pure white, half-white, quarter-white, cream. Collecting varieties of linen had become a genteel mania for the Dutch, who gave cloth woven of flax the veneration the Chinese gave silk. Holland produced extraordinary flax; its blossoms grew the deepest blue of all flax flowers in the world, its stems the richest yellow of all flax stems. Weavers passed along the treasured secrets of a supertight warp and weft. Bleach girls soaked freshly woven, field-long strips of cloth in whey and laid them out on fields of grass under the clear, clean skies of Haarlem. The town's climatic conditions gained such fame for their whitening properties that the swells of Paris sent their laundry all the way across the continent for freshening.

In Europe, the tablecloth had been held supreme among linen products since medieval times, when only royalty could afford large sheets of fabric woven of flax. Pressing mealtime accoutrements had itself become an art, with tablecloths ironed in crisp checkerboard patterns and napkins folded by professionals in the shape of artichokes or suckling pigs. Personal estate records in the New World attest to the continuing Dutch passion for linen, though the emphasis in America was decidedly practical rather than whimsical. One colonist's far-from-atypical inventory, for example, included thirty-two pillowcases and 103 bedsheets—23 linen and 80 muslin. Each shallow drawer of Margaret's *kas* would be lined with a sheet of clean linen, which she also would use to top its contents: wadded bunches of petticoats and other full skirts, smoothed bedsheets and napkins, hand-towels and nose-towels (we know them as handkerchiefs). Perhaps most important were the rolled bolts of diaper, a textile put to use for everything from rags to washcloths to swaddling blankets and baby clothes. Over the years to come, Margaret's sleek Stone Street palace would become a household that bustled with children—she would deliver another girl and two boys by 1670—and every yard of plain, utilitarian diaper would come in handy.

7

The Superior Authority Over Both Ship and Cargo

On August 26, 1664, four British frigates under the command of Colonel Richard Nicolls entered the waters off Manhattan, weighing anchor in Gravesend Bay, at Breukelen, opposite New Amsterdam. An expeditionary force of four hundred soldiers and sailors disembarked and captured the ferry there about as easily as you'd take away a child's doll. For a week the fleet sat just across the East River from Manhattan, close enough for Margaret to get a good look at the men's artillery from the second-story window of her Stone Street home. On September 8, less than two weeks later, Director-General Peter Stuyvesant signed a surrender presented by Colonel Nicolls, making New Netherland into New York. Not a shot was fired.

Scholars of New York history have long debated the reasons why the Dutch colonists went so meekly. They could have fought harder to preserve Dutch dominion. They could have fought at all! Across the Atlantic, leaders of the Dutch Republic were outraged by the British invasion—especially because the United Provinces and England had signed an oath of official friendship just two years earlier. Dutch officials also were incredulous that the settlers did not fight for their colony.

There were actually a few good reasons for the colonists to cede power so readily. They were militarily outnumbered: The Dutch could count only 150 troops, matched against a rumored 2,000 British fighting men packed into those warships waiting at Manhattan's watery doorstep. Even if the city raised every able-bodied man who could possibly bear arms, there were no more than 250 or so on the island. Also, the arsenal at New Amsterdam clearly paled in comparison to the firepower on the ships across the East River. Supplies of powder were pitifully low, six hundred pounds tops, depleted by the recent Esopus war. Fort Amsterdam boasted an unimpressive twenty-four pieces of artillery against almost three times as many English guns. The planks of Stuyvesant's wall at the northern edge of the city had rotted (though it would be decades before they fell). Manhattan's riverbanks had no fortification whatsoever. The city's stores of provisions could never sustain its people through any kind of prolonged siege.

And England was more determined than ever to claim New Netherland as its own. Not willing to wait for the Dutch to acquiesce, King Charles II had already awarded the Dutch province to his "Dearest Brother" James, Duke of York and Lord High Admiral, basing the English claim on the idea that Henry Hudson had been "an English gentleman" who had erroneously sold his navigational "maps and cards" to the Dutch.

A political prodigy of just past thirty, brother James stood to personally gain from the gift, though the colony seemed to offer paltry remuneration in comparison with his other pet projects, including the Company of Royal Adventurers Trading to Africa and the Morocco Company, both of which were attempts to loosen the West India Company's vise grip on the slave trade. New Netherland would earn the duke up to thirty thousand pounds a year in customs duties and rents—not bad, but nothing like slaving revenues.

The territory, however, did offer something quite incredible—the Hudson River. The corridor from the Atlantic all the way to Montreal would grant England the military and commercial base it needed against the French in Canada. The colony also made a vital link between the seaboard towns of the east coast. And it was already producing food in some abundance—beef, peas, biscuit, and rum—that the sugar planters of the British West Indies needed to provision their estates. All the Duke of York had to do to reap these advantages was give his brother Charles an annual payment of forty beaver skins.

There was just one roadblock to immediate surrender, and that was Peter Stuyvesant.

At first, he seemed blithely unaware of the imminent threat. The English and Dutch had continued to wrangle over territory since the last threatened invasion; that was business as usual, especially as British holdings and colonists increased. Both the Company directors in Holland and Stuyvesant had heard rumors of a British fleet approaching back in July, but King Charles's commissioners successfully smoothed the feathers of the directors and through them Stuyvesant with a tale that the fleet was coming to America only to "consolidate the existing English provinces" or "install the Episcopal government as in Old England."

The fabrication worked. Stuyvesant left the colony to visit Fort Orange on August 6 and help negotiate a peace treaty between the Mohawk and the Onakouques. During his three weeks of travel, the director-general fell ill with the era's pernicious, untreatable fever. He only dragged himself back to Manhattan after his council sent an urgent message that merchant ships had spotted a British fleet headed for New Amsterdam. Stuyvesant arrived home to find a work party of slaves and burghers doing their best to bolster the fort. But only a fool would think such efforts could shore up the island's defenses. So it must have been with a certain miasmic horror that he straggled from his sickbed on August 26, the morning after his return, to confront the unthinkable. Colonel Nicolls had stepped ashore at Gravesend to proclaim the Duke of York's undisputable authority over all of Long Island.

Both Colonel Nicolls and Governor John Winthrop, Jr., of Connecticut sent Stuyvesant letters demanding surrender. He hid one letter and tore the other to pieces. The colonists demanded that the scraps be glued together. There were no takers for Stuyvesant's campaign of resistance. The residents of Long Island had been the first to decline his invitation to fight, then the troops at Fort Orange, and finally the residents of New Amsterdam. Even his seventeen-year-old son Balthazar tried to talk Stuyvesant out of resisting the occupation, no matter how he despised the English. Ninety-three of the most powerful men of New Amsterdam presented the director with a signed petition: "We . . . cannot consci-entiously foresee," it read, "that anything else is to be expected for this fort and city of Manhattans [. . .] than misery, sorrow, conflagration, the dishonor of women, murder of children in their cradles, and, in a word, the absolute ruin and destruction of about fifteen hundred innocent souls."

Neither his fever nor his wooden leg nor a lack of backup stopped Stuyvesant from scaling the ramparts of Fort Amsterdam on September 4, when he spied a detachment of two British ships slowly cruising toward the tip of Manhattan. Atop the bastion of the fort he threatened to fire the first cannon, even if it meant fighting this war single-handedly.

It seemed Stuyvesant was the only one to prefer probable failure over certain survival.

If the British stormed Manhattan, colonists knew, it would spell suicide. The English squadron had departed Brooklyn to sail up the

Hudson and park itself athwart Governors Island, effectively block-
ing any escape through the Narrows. Then a merchant ship flying
the English colors appeared. Eager soldiers from New England and
from English-dominated villages on Long Island arrived every day
to join the British. The English citizens of New Amsterdam had
abandoned the Dutch. Fort Amsterdam's walls were both flimsy
and highly flammable; the structure was too small to provide shelter for all the island's twelve hundred women and children. The
city had just two dozen barrels of drinking water and barely enough
grain to sustain the populace. Finally, the most seasoned Dutch
Reformed minister climbed alongside the director-general in order
to subdue Stuyvesant's one-man campaign atop Fort Amsterdam.

Still, the question remains: Why after half a century of Dutch
investment in these lands did the citizens of New Netherland refuse to join Stuyvesant and put up a fight? Why not defend their
honor as Hollanders?

With their lives in the balance and harbor traffic blockaded,
merchants were hemorrhaging money every day the standoff continued. The slave ship *Gideon* was a visible symbol of the damage
to Manhattan's trade. The ship happened to arrive in the East River
harbor from Loango by way of Curaçao just days before His
Majesty's ships anchored at Gravesend. The stench emanating from
the *Gideon*'s hold—which held a cargo of 290 men and women in
various stages of deprivation, sickness, and starvation—grew more
foul every hour that the English-Dutch conflict remained unre-
solved. As long as the Dutch stood off against the English, the ship's
owner could not distribute his merchandise and collect his profit.
His "product" differed from the tobacco and linens Margaret's ships
carried, but the hardship posed by this paralysis of maritime trade
was exactly the same. An actual war might halt trade indefinitely.

A more subtle argument could be made that trade under the
British might be less onerous than it had become for the Dutch.

This was certainly true for some of the most influential merchants, such as the Hardenbroeck-Philipse franchise. England's most recent Navigation Act, passed by Parliament upon Charles II's accession to the throne, was aimed at booting the Dutch out of the American colonial trade entirely. It mandated that sugar, tobacco, and indigo must be shipped to England only in English ships, and that all merchandise bound for British colonies first must pass through British ports. Perhaps, reasoned Dutch merchants, coming under British rule would offer some leeway around these regulations.

Switching to foreign rule obviously would not be easy. But colonists were mollified by the language of the surrender documents Stuyvesant signed four days after his quixotic stand. The Articles of Capitulation imposed few hardships on the Dutch, and instead offered protection of the rights and traditions they had known in the past. The document's twenty-three points decreed, among other things, that there would be no expropriation of Dutch property and no punitive expulsion of colonists. Dutch settlers were free to stay or leave. If they stayed, they could keep their houses, land, goods, and ships—provided they took an oath of loyalty to the English king. Dutch inheritance customs would remain as they were. Contracts would be enforced "according to the manner of the Dutch." So would educational practices, religious observances, and the language spoken in the province. The Reformed Church would continue to collect the taxes and run the schools. Services would be held in Dutch, as always.

But the transfer of New Netherland was not a done deal, despite Stuyvesant's signature. Aggravated over losing their colony, Holland's leaders instigated war against England. Three years of spirited naval engagements spanned the Channel and also broke out along the west coast of Africa. The pristine slice of New World territory that comprised New Netherland never hosted any combat; after relinquishing the province, colonists behaved as though English rule was a fact of life and did not fight it. (With one exception, in 1673, when Holland would briefly reassert its claim over New York with a bloodless assault that resulted in a heady six months of control over the province.)

While New Netherland refused to fight, Holland battled on. The 1664 war that had begun on Manhattan officially ended only after Dutch ships assailed the Thames, burning three men-of-war and capturing the *Royal Charles*—all this in the grim shadow of the Great Fire that devastated London the previous year, and a bout of plague that took twenty-six thousand lives. In July 1667 Charles II sued for peace. It was yet another Dutch victory on the high seas.

Then a strange thing happened. The Dutch had won the war they brought to avenge the invasion of New Netherland. The time had come to firmly request the return of that colony. Instead, the West India Company's directors claimed as its prize a territory called Surinam, a tiny clot of jungle on the northeastern "wild coast" of South America that happened to belong to King Charles. (And while they were at it, wrangled the Dutch Republic, England could throw in the 174-mile-square isle of Curaçao.) Let King Charles take Manhattan.

To someone who had sunk a fair amount of money or labor or even affection into New Netherland, the irrevocable transfer of its lands to a long-sworn enemy of Holland must have come as a shock. But at the time, the sacrifice of New York for Surinam had a mighty logic.

Ten years before, the Company had lost its lucrative Brazilian sugar outpost to the Portuguese. The salable treasures of North America had expanded the coffers of private traders far more than those of the Company's investors. In 1667 Company directors saw the African slave trade—centered around sugar plantations in South America and the Caribbean islands—as a most attractive profit-generating venue. The way the directors construed it, if they did not choose Door A, which happened to be Surinam, then the Company and their personal fortunes would almost certainly fail. Door B, New Netherland, would never produce a profitable slave trade. The province was too far from the source, so ships' cargoes had a tendency to become less valuable (i.e., perish) en route. Furthermore, the directors predicted, there would never be as great a market for slaves among the farmers of North America as among the agri-kingdoms of the Indies.

Despite its new rulers, the colony remained Dutch in ambience for a decade. Colonel Richard Nicolls, the first governor of British New York, immediately sought to curry favor with Margaret and Frederick and their big-money peers in the colony. Governor Nicolls deemed New York City his kind of town: "The best," he informed the Duke of York, "of all his Majesty's towns in America." The governor went to bat for the merchants with whom he socialized, pleading to exempt New York from the strictures of the Navigation Acts. Overly binding commercial regulations, he knew, would cut into the fur trade. Traders such as Margaret needed full holds of European goods to barter with the natives for pelts.

Nicolls even befriended his former adversary Stuyvesant, who had returned from Holland when the coast cleared to gentleman-farm his sprawling *boewerie,* which grew to be as famous a show-piece as his Great House had once been. Stuyvesant turned over his harborside mansion to the British conquerors, who renamed it Whitehall after the English royal palace in London; the Dutch fort became Fort James.

The Dutch were not held in uniform respect; some British citizens sneered at their customs and referred to them as butter-boxes, milk-and-cheese men. Official lenience was also short-lived:

Parliament soon decreed that merchandise from abroad must come either to or through England, in English ships, then, as time went on, only with an English captain, with an English crew.

But these strictures backfired. The English merchants did not have the European contacts to get the merchandise they needed to trade in New York. Only the Dutch, experienced traders such as Margaret Hardenbroeck, could procure the fine fabrics, stockings, bricks, looking glasses, and even the mellow cheeses the Old World still provided for the New. As a result, Governor Nicolls made an appeal to the English powers, which Peter Stuyvesant helpfully seconded as an adviser. (The Gouda pipeline must not be stoppered!) The British granted a reprieve: The merchants of Holland could ship their wares freely for seven years, but would be limited to three vessels per year.

Yet once again the lenience was brief, and Margaret unexpectedly found herself at the center of the most crucial commercial controversy of the day.

At the end of 1668 Margaret was in Amsterdam on one of her regular trade missions when she received bad news: The British had changed their minds and reduced the Dutch ship allowance, this time from three ships a year to just one. After early conciliatory gestures, the English were moving to enforce their primacy in international commerce. The policy posed an immediate and potentially costly problem for the thirty-one-year-old merchant. She was working with sixteen other international merchants to ready her ship, the *King Charles,* for a return trip to America. Unfortunately, Margaret already had another ship well under way. Given the new rules, her *King Charles,* fully laden with cargo and crew, would have to sit at the dock until the next year to be eligible to sail. Margaret was stuck in Amsterdam. And even if her ship sailed in January, the soonest allowed, the timing would put ship, crew, and cargo at risk from deadly winter sailing conditions. Jointly with her peers, Margaret drafted a petition to "Your sacred Majesty" pleading for an exception to the new rule. She asked the king to take the merchants' "ruinous condition" into his "princely consideration, upon which depends ye Welfare or Destruction at once of us, our Wives and Children," and authorize the departure of the paralyzed vessel. On December 11, 1668, the king granted special permission for her ship to make that one voyage.

For his part, Margaret's husband had wasted no time in taking the oath of loyalty to the Crown. Frederick pledged to obey the Duke of York and his governors, though none of the Dutch colonists would be actual English citizens. There is no record of Margaret swearing an oath, which is not surprising. Under British law her husband would be expected to represent her, making his allegiance her allegiance. That assertion of loyalty bought the couple and other Dutch merchants favorable treatment in the years after the British conquest.

Margaret and Frederick took advantage of the early English desire to placate the Dutch and appealed to Governor Nicolls for permission to conduct business in the town of Albany—a savvy move. With the declining availability of skins in the lower Hudson Valley, they needed to expand beyond Manhattan to procure their principal product. To the north and west, the forests still held considerable beaver and other game, and Indians were eager to engage in trapping them for barter. Accordingly, the governor granted the couple a pass to traffic directly with the Indians. They could then either purchase pelts from the merchants who dealt with the natives in the Albany market or cut out the middlemen and go to the trappers directly.

Changes in currency valuation further enriched the couple. Margaret and Frederick had invested in wampum at just the right moment; records show that in the early 1660s they had laid in hogsheads full of the more valuable black beads from the purple quahog, as well as the white beads. When the English took over, the wampum supply shrank and its value increased by 400 percent. Margaret and Frederick began to stash large quantities of wampum in anticipation of an even greater hike in value, which took place in 1666.

*T*he British haute style showed itself socially as well as politically early on when Governor Nicolls, a horse fanatic, established a racecourse called Salisbury Plain in Hempstead, Long Island. The track was a frothy digression from the rougher Dutch amusements (gin drinking, kitten bashing). Nicolls purportedly arrived to open the season in a crested chariot, with a long ostrich plume falling from his hat over elaborate tendrils of long, perfectly coifed hair. Charles II himself could not have been more dashing. The governor's retinue showcased ladies in skyscraper hairdos and ruffled, bosom-bursting gowns. These minions stood out even among the most well-appointed Dutch.

With Salisbury Plain, social New York had officially been christened. And greater sophistication was joined by a closer connection with the outside world. When Francis Lovelace followed Nicolls as governor, he instituted the first regular postal service in the province, based at the foot of Broad Street, hard by the old burgher bridge. Merchants still congregated there to scan the harbor and trade information, but now they had an official mercantile exchange. The Heerengracht canal was paved over in 1675 by order of Lovelace's successor, Edmund Andros. A new stone pier, the Great Dock, cleaved the waters of the East River, and offered more protection from the elements than ships anchoring at Manhattan had ever known.

That older Dutch ways still prevailed is suggested by records of the post-British existence of Margaret's brother Abel, who continued to skulk around the community's rougher edges. In summer of 1665, one Denys Isaackson drew a knife on Abel Hardenbroeck as Abel sauntered with "some women folk towards the Bouwery" in the vicinity of present-day Astor Place. Abel's own apprentice fled in 1670, complaining of beatings and mistreatment at his master's hands. Hardenbroeck thought it appropriate to sue the abused boy's mother, demanding that she return her son to him and "pay for loss of time" as well. All the slickness of the British could not immediately erase the earthy ways of the Dutch.

Like Nicolls, his two successors sought to cultivate Dutch merchants despite official restrictions. Lovelace's deputy governor profited directly from an investment in a Hardenbroeck-Philipse–owned ship, the *Duke of York*. Governor Andros, appointed in 1674, enacted new regulations that helped increase the volume of shipping in the city ten times over during his term. Andros decreed that Manhattan would be the only site where imported goods could enter the colony. Likewise, ships could only load their cargoes for export out of New York Harbor. No one outside the city could legally pack wheat, beef, or pork to ship abroad.

In 1675 Andros invited Frederick to sit on his advisory council. The choice was politic: In 1674, after a little over a decade of marriage to Margaret and steady investment with her in furs, European goods, wampum, and real estate, Frederick Philipse was officially listed as the richest man in New York. His net worth was eighty thousand guilders, far surpassing the wealth of his nearest rival, whose estate was valued at fifty thousand. Given Margaret's integral part in building that fortune, Frederick's identification as the richest man in the colony certainly implicates her as its richest woman. In keeping with British law and custom, however, there was no similar measurement of Margaret's wealth.

The omission was significant. The invasion had imposed England's legal system on the colony. In starkest terms, that meant the gradual elimination of Manhattan's independent she-merchants and, more broadly, greatly reduced status and legal standing for every woman in New York for generations to come. Women with commercial interests could no longer legally sign contracts, represent themselves in court, or otherwise take a public role in managing a business. For a time, Margaret and other businesswomen simply ignored the new rules. But that would not work forever.

The option of marriage under *usus* also perished. Nuptial laws changed gradually—perhaps to hold off any protest—but by the mid-1670s every married woman was a legal entity under the protection of her husband, just like the Dutch woman under *manus*. A prenuptial contract spelling out a woman's independent legal standing was no longer an option, at least officially (the Dutch still abided by their customs no matter what laws the British brought). Inheritance law had changed even earlier, in 1664; the Duke's Laws stated that the eldest son of the deceased would receive a double portion of money or land. No longer would it be a matter of course for Dutch families to divide an estate among male and female offspring equally. When a woman's husband died, she could get back her dower, but she would be forced out of her husband's house after forty days while the court assigned her legally mandated third of his estate. Joint wills between husbands and wives declined. Even the practice of retaining a woman's surname after marriage would grow extinct.

And so Margaret had to adjust, to make her way around the edges of the system. To be flexible. To be canny. To compromise when necessary. She retained the power of attorney for Frederick when it served the couple's ends and managed her shipping concern as she always had. But now she had to move more carefully, as her own status changed and the rules for trade grew more onerous.

As of 1674, the Crown collected duties on all goods en route from Amsterdam to America not only in England but also in New York. In 1677, as a mature woman of forty and a Manhattan resident for almost two decades, Margaret sought to get around the rules again, this time with a direct appeal to the Duke of York. She sent a letter to England requesting permission to purchase a Dutch ship "in hopes to make her free," in other words with the expectation that authorities would declare it an English ship and thus excused from the excessive taxation. The duke's secretary sent a curt summary of the incident to New York governor Andros; his phrasing suggests a certain discomfort with the she-merchant's methods. He "diswaded her from it all I could," the secretary wrote, "by reason of ye strict orders of late prohibiting any of those practices heretofore."

By the end of the 1670s the fur business had begun to diminish for the smaller traders, but not for those with good connections to the source and material goods to offer them. At Leipzig, the premier fur market of Western Europe, or even at the smaller Amsterdam *kermis* fair in September, merchants who could obtain large quantities of what was called "good merchantable beaver" reaped enormous profits. Margaret and Frederick's trade fleet had dramatically expanded. Each spouse had a namesake ship: the *Margaret* and the *Frederick* typically transported tobacco and timber as well as hides to Amsterdam. Surviving records regarding ship ownership are not complete, but port documents suggest that their fleet included as many as two dozen ships at one time or another, including part ownerships: the *Charles, Pearl, Beaver, New Netherland Indian, Morning Star, Charity, Delaware Merchant, New York Merchant, Dolphin, Fortune, Francis and Thomas, Hopewell, Jacob, John & Rebecca, New York, Pembroke, Philip, Unity, Katharine, Vine, King Charles,* and *Ketch Royal.*

As other fur traders slipped beneath the commercial waves, Margaret and Frederick continued to expand their enterprises. A typical bill of lading for the *Beaver* offers an impressive picture of their inventory. The ship cleared the port carrying 1,713 beaver pelts, 502 bear, 1,550 buck, 46 cowhides, 1,250 standard foxes and

53 timber grays, 94 timber minks, 1,100 "catts," 1,100 otters, 363 "musquashes," and others. Returning to New York, the same ship brought linen, 3,537 ells of it, along with 72 gross tobacco pipes, 45 swords, and 99 musket barrels, plus smaller quantities of tools, books, and woolen nightcaps. Another cargo the same year helped ensure that New Yorkers would satisfy their need for cooking pots, stockings, nails, thread, and Holland cheese.

Yet even as Margaret was experiencing uncommon commercial success, her personal life seems to have taken a rockier turn. Evidence of frictions in her marriage surfaces in a single highly suggestive source—a terse legal document from 1675 that gives a rare glimpse behind the conventional split Dutch doors into the life of a strong-willed thirty-seven-year-old at a moment of domestic turmoil.

The document, a transcription of minutes from an Albany courthouse hearing, describes a case that began with a visit paid to the Hardenbroeck-Philipse lodgings one summer day several weeks earlier by two proper Dutch ministers, Dominie Gideon Schaets and Dominie Nicholaes van Rensselaer. All was not well, it seems, in the house of Margaret and Frederick. For reasons that are left unexamined in the court document, Margaret had apparently upset the marital status quo by declining to accompany her husband to New York City as he wished.

That this was no trivial domestic spat is indicated by Frederick's next course of action: He took himself over to Albany's Dutch Reformed Church to plead the intercession of the dominies. As the tight-knit frontier community's de facto social workers, its religious leaders were often called in to provide family counseling. Perhaps, Frederick hoped, they would be able to talk sense into Margaret. The court minutes capture the uncharacteristic neediness of the phenomenally successful merchant, now nearly fifty years of age, whose public power seems to have had little currency in the marital arena. Would it be possible, asked Frederick, for the dominies to go to his house and persuade his wife "if possible with kind words" to reconcile with him?

The dominies would try. Margaret's greeting, however, must have told them it wouldn't be easy. "You have the devil in you!" she offered as Schaets and Van Rensselaer crossed her threshold. Was this intentional blasphemy, or was Margaret just extremely irritated at the churchmen's intrusion into her domain, physical and marital? Most likely the latter. She was, after all, so loyal an adherent as to have pledged funds to strengthen the church infrastructure in Bergen. Now Margaret had made a decision. It was a private issue, and her choice to make. Couldn't they all leave her be?

Margaret's insult did not endear her to the ministers. Yet it was the events that followed that led ultimately to the charge characterized in the court minutes, brought by Van Rensselaer, for the use of "very prophane and godless language." And here is where the court record introduces an unexpected piece of the domestic puzzle, in the form of an individual named Jan Gerritson, the subject of the legal complaint, who was present with the rest in Margaret and Frederick's house that afternoon.

Gerritson, described as "their servant," comes across in the record as anything but servile. He burst into the room "in an angry and excited mood . . . uttering bitter words" and went further than Margaret had in cursing the ministers, even accusing the two men of having been responsible for alienating Margaret from her husband to begin with. That contention seems to have pushed Frederick, "his master," over the edge, since he is described at that point as brusquely taking Gerritson "by the arm and thrust[ing] him through the inner door."

To the presence of stubborn Frederick (recall the Wiltcyck wheat debacle) and determined Margaret, we now add an impassioned third party who was apparently driven by some brand of spiritual fanaticism. The ministers were false prophets, Jan announced, while he himself was "imbued with the Spirit of God." He seems to have been drunk as well (he admitted later to imbibing "a glass of wine" before accosting the dominies). The two churchmen added their own flavorful personalities to the proceedings. Van Rensselaer was a novice minister installed through political connections; his reputation as a half-mad mystic would get him dismissed after only a year. He never got along with Schaets, a hard-drinking fixture of the town who was famous for his vituperative attacks on anyone outside the fold of Reformed Church orthodoxy.

With this crew, Frederick would not be taking Margaret off to New York anytime soon.

Perhaps the most curious aspect of the 1675 court transcript is the rapport between Margaret and Gerritson. Jan comes across as more than a little attached to the idea that the she-merchant is in some way his soul mate. He makes it his charge to protect Margaret, gruffly keeping the ministers and her husband at bay. The following day, according to the record, Schaets and Van Rensselaer reentered the Philipse dwelling in search of Gerritson (who had not, apparently, been driven off by Frederick). Pressed by Van Rensselaer, the servant refused to repent. Then he offered an analysis of the household drama that surely made Frederick wince. Margaret, the man contended, "clung to him, Jan Gerritson, as her God and Saviour."

Did Margaret seek out Jan for spiritual sustenance, given his claim to be "the true prophet"? Or perhaps the closeness Gerritson alluded to was emotional, born of Margaret's need for solace in the growing disunion between her and her husband. Perhaps she even "clung to him" physically. The servant lived intimately with the couple, sleeping under their roof. It would seem easy to stray if you were a woman gifted with a great deal of autonomy, an equally independent husband, and a willing man on the premises.

But would a churchgoing Dutch-born woman have engaged in such an affair? Even if her marriage had taken a troubled turn?

Through Dutch history, a premium had been placed on marital harmony. All society expected wives and husbands who experienced difficult times to make peace and move on. Adultery was one of the few sins the Reformed Church deemed serious enough to warrant divorce. Still, this was the New World, whose denizens distinguished themselves by the excesses of their comportment. Margaret may have differed from her peers only in the sense that if she wanted to defy her husband's wishes, live apart from him on occasion, take a lover, and earn the acrimony of her Church, she would have had few compunctions about it.

The 1675 transcript ends with defendant Gerritson escorted to Manhattan "to defend himself there before the general Court of Assizes." It seems unlikely that Margaret would have accompanied Jan to lend her support. Her business in Albany was thriving, after all, and it was still high season for trading furs. When the authorities released her servant, he would know where to find her.

*M*argaret's marriage continued. She and Frederick remained business partners. Their work, of course, afforded the couple a great deal of distance from each other on occasion.

Even so, there were the children to raise. By 1679 Margaret had five children under the age of twenty-one. There was Eva, now nineteen; Philip, sixteen; Adolph, fourteen; Annetje, twelve; and nine-year-old Rombout. Sometimes she left her children at home when she traveled, whether she sailed up the river to trade for a few days or made a months-long overseas venture to Amsterdam or London. Her husband, parents, and servants were available to care for them. Sometimes Margaret took one of the children with her over the ocean. Given the typical conditions of transatlantic travel on any merchant ship, Margaret was clearly not one to pamper her offspring.

Annetje made a cameo appearance in a narrative that chronicled one of her mother's voyages. Read more than three centuries later, it provides a uniquely vivid, if somewhat harsh, picture of the forty-two-year-old Margaret engrossed in the activity that was central to her life.

Passenger Jasper Danckaerts, a young man from Holland, kept a diary during a six-week crossing to New York that he undertook in one of Margaret's principal ships. Margaret receives prime billing (under her married name) in Danckaerts's tale, *Journal of a Voyage to New York*. His title page describes the voyage as undertaken:

> In the small Flute-ship, called the *Charles,* of which Thomas Singleton was Master; but the superior Authority over both Ship and Cargo was Margaret Filipse, who was the Owner of both . . .

Danckaerts and colleague Peter Sluyter were missionaries scouring America for a hospitable spot to relocate their sect, a fringe group of Christians known as Labadists, who were no longer welcome in Holland or elsewhere in Europe and had set up a community in Surinam. While the Labadists' views surely colored the picture Danckaerts draws of Margaret—she was anything but the demure, sober female the sect celebrated—his written portrayal went unpublished and never reached her peers in her lifetime. But Danckaerts and Sluyter were nothing if not gossips, proselytizing their way around Manhattan, out to Long Island, up the Hudson Valley corridor in the months following the voyage. It would not be surprising if Margaret's local reputation took a beating, because Danckaerts was as opinionated as he was fanatical.

A workhorse of a flute-ship, the *Charles* set its course from Amsterdam in June 1679, onloading provisions and passengers both there and at the Texel, the outlying port from Amsterdam where transatlantic ships stopped briefly on their way out to sea. An account of its cargo, made at the shipping office in Plymouth, England, lists eighty-eight different types of goods on board. This staggering variety underscores not only the domestic needs of New York consumers but also the attention that Margaret as supercargo—a job that entailed overseeing all merchandise on board—had to put toward arranging to ship all this diverse merchandise.

Dry goods packed much of the ship's hold. The most valuable item on board was a 2,872-ell length of Holland linen taxed at more than thirty-five pounds. The ship held other textiles as well, including 10,045 ells of "ozenbrigs," a coarse all-purpose fabric with a duty of twenty pounds, and smaller quantities of diaper, "hamburgh linen," calico, damask, duck, holland, lawn, silk, and "Smirna Demity." Margaret had invested in women's and girls' bodices, mittens, stockings, thread, buttons, "sissors," twine, knives, and sword blades. In the capacious bowels of the ship—traders preferred fluteships specifically for the amount of merchandise they could squeeze in—containers of "ffrying pans" were jammed up against "Alcemy Spoones," chimney bricks nestled against "Dutch Printd Bookes," five dozen bridle bits shared space with "a pcell [parcell] Pewter toyes for Children." (Perhaps her own?) And that is not nearly half the lading.

Along with the captain and ten crew members, the ship carried a dozen passengers who had paid for berths from Holland to England and all the way to New York. Jasper Danckaerts introduced himself to Margaret as J. Shilders and his companion as P. Vorstman, identities they had invented with the idea that they must pursue their mission on the hush.

Perhaps she sensed the men's fakery. But it was an unnecessary dodge. Margaret, despite her minister-directed outburst in Albany four years before, did not have anything against religious worship. Her family, like the rest of the Dutch community on Manhattan, attended the Reformed Church. She had given money to bring a dominie to Bergen. That gesture went, of course, to shore up her church on the godless frontier. But Margaret had no beef with other faiths, a tolerance she shared with the majority of her fellow colonists, who approached the pluralistic nature of their community with practicality as much as philosophical largesse. The unofficial religion of New York, then as now, was making money.

Religious or cultural differences could all be very nicely subsumed under that overarching goal. Members of the Jewish faith, for example, had found a refuge on Dutch Manhattan early on, even when Peter Stuyvesant himself made it clear that he did not want to host any "blasphemers of Christ" in his colony. But when he proposed barring Jews, the Company advised him to let them "peacefully carry on their business" and to treat them "quietly and leniently." This made practical sense: The community of Jewish merchants in Amsterdam had invested considerable capital in the Company, first of all, and the province needed every able hand it could get. It was not immediate, but New York Jews gradually gained respect for their observance as they did for their business practices. The Philipse family shared shipping projects with Jewish merchants over the years because they were commercially compatible.

Even Quakers—so despised as heretics when the first group of refugees from England disembarked their ship on the Strand at New Amsterdam that the women were immediately dragged off to jail by their hair—now enjoyed tolerance. The neighbors of a group of Friends who settled in Flushing would have none of Stuyvesant's efforts to banish them, advising that "We desire . . . in this case not to judge least we be judged, neither to condemn least we be condemned, but rather [to] let every man stand and fall to his own Master. We are bound by the law to do good unto all men."

Margaret's approach showed itself in her choice of captain, Thomas Singleton, who was a Quaker. The supercargo ordinarily equaled the rank of the captain, but in this case Margaret outranked Singleton because she also owned the ship. When Singleton furnished the proud young wife who accompanied him across the sea with the wherewithal for plenty of silver and gold baubles, a disapproving Margaret demanded, "Why did you give her the money to buy them?" But though she might have frowned on his extravagance she would not discriminate against Singleton due to his faith; she simply hired the best skipper available.

It must have been annoying to find Danckaerts posing nosy questions every time she turned around. He and his traveling companion would draw her into a discourse on the deck of the ship and then scamper off to their bunks to dip their quills and record the conversation. Danckaerts had plenty of opportunity to watch Margaret run the ship—and to criticize her. He judged her time and time again as guilty of "terrible parsimony," "miserable covetousness," and, above all, arrogance. He found her style of authority as abrasive as it was decisive.

Any incident would set him off. There was the time a girl lost the mop she was rinsing over the side of the ship. From Margaret's perspective, it made sense to launch the jolly boat to retrieve the swab before it floated away; they had traveled some weeks from Holland and had weeks more until they reached England, and it would be difficult on the open seas to replace it. Danckaerts maintained that the mop was worth only six cents, and there was too great a risk to send out the little boat with two valuable sailors, who returned chilled and exhausted from the labor of rowing over the swells. But rescuing the swab was what Margaret knew. This was how she earned her money, conserving shilling after shilling. After years of running her business that way the shillings unquestionably piled up.

Much to Danckaert's horror, Margaret had even taken an intimate. He was her right-hand man aboard the ship, an intense, impulsive Dutchman identified only as Jan. In Danckaert's description, Jan appears as Margaret's closest companion, reporting to her directly and yet more powerful than the average servant; they shared an open-secret liaison that might have been born of proximity in the cramped quarters of the *Charles*—out in the middle of the lonely ocean, far from the proprieties of shore—but more likely had its origins inland, in a rented Albany townhouse. The captain, Danckaerts writes, "was very assiduous or officious to please [. . .], especially Margaret and her man." Margaret's other man, Frederick Philipse, had stayed behind for this trip, as he often did; when a couple dealt in as much merchandise as these two, they might have told people, they could not stay side-by-side 365 days a year.

Jan, as described by Danckaerts, bears an unmistakable resemblance to the Jan Gerritson of Albany whose profane bluster earned him a ticket to Manhattan to stand trial in the summer of 1765. On the voyage of the *Charles,* it was Jan who broke out Margaret's best Madeira on a whim, without awaiting permission, and poured a round or two for the officers' mess. In Danckaerts's view Jan was "the greatest grumbler" and expressed himself with profanity "worse then the foulest sailor." He is quoted as making comments about Margaret herself, the personal tone of which suggest some kind of intimacy between him and his mistress. Awaiting the arrival of Margaret at the start of the voyage, and afraid her delay would mean missing a favorable wind, Jan had this to say: "If this wind blows over our heads, I will write her a letter which will make her ears tingle." However brusque, however rude he was, Margaret seemed unfazed by his behavior, and though he was not a member of the crew, nor a mariner, she ranked Jan's opinion about nautical matters alongside that of the captain.

Still, Jan did not rule the ship. Margaret's word was law. She had responsibility for the survival of all on board. When they landed in port, she arranged the loading and unloading of all cargo, and issued orders to sailors slacking off at the windlass. Though the crew reported to the skipper, Margaret actively oversaw all that transpired, everywhere on board, every day. She had responsibility for Annetje too. Margaret's twelve-year-old minded herself admirably, spending most of the voyage scrambling into the corners of the ship, spying on the crew, and staying out of the way of her mother.

At Falmouth, Margaret disembarked to go on foot into town, then supervised the off-loading of merchandise destined for another of her ships, a larger one bound for Barbados. The missionaries remained in the port writing letters, hanging around the ship. They understood no English (Margaret had long since grown fluent) and only disembarked to attend church services at every stop in England, gathering material on the heathen.

Even Margaret's generous gestures brought forth the judgment of the Labadists. As the *Charles* approached Virginia, nearly home but running short of supplies, a ship on its way to England approached. The skipper crossed in his jolly boat to pay a call. He described the storms and contrary winds he had encountered, saying that many of the crew had fallen sick. Margaret sold the captain a hogshead of the ship's beer. She received nothing but dark looks from her own crew, especially since the skipper afterward presented little Annetje with "a good lump of gold" and some apples. Danckaerts was incensed: So the people on her ship go without and her little girl gets a gift?

Margaret was unfazed. She offered a ration of beer three times daily. She hired the cook to prepare basic meals, which were decent, at least for the cabin diners. She had traveled overseas for more than twenty years, and she knew the privations. Passengers were free to bring further supplies on board. They could share whatever fresh catch they obtained, or lay in brandy or fresh butter from the countryside whenever the ship docked en route.

Danckaerts's journal shows just how terrifying it could be to cross the Atlantic on a small merchant ship and how challenging it must have been for Margaret to bear ultimate responsibility for the vessel and all its goods and passengers. The stakes were as high as a murderous, cresting wave in an ocean storm. Just before departing England for North America, Margaret received word that Turkish marauders had seized four Dutch ships. Soon after, a privateer's ship of war appeared and disappeared in shreds of mist. No one knew if the flute-ship could outrun this ghostly vessel. Panic overcame the passengers. With that danger only just past, hard rains came. A cold wind tossed the ship for days until the moaning of the women made a terrible music on deck. There was nothing to eat toward the end of the voyage but the heads of salt fish, just about spoiled. The men stayed drunk.

Yet the journey of the *Charles* was a successful one. All of Margaret's merchandise arrived safely, as did her passengers, and the ship would go on to many more crossings. As improbable as it was that a seventeenth-century woman in America would triumph over such obstacles, the job comprised the ordinary stuff of Margaret's life. She knew the sea and she knew her business. Only now, Margaret paid less mind to the sights and sounds of the crossing than she did in her youth. She hardly glanced at the blue dove above the pale green waves that flew to the ship as if to welcome it back to America.

*F*rom the balcony of her rooftop on Stone Street, Margaret held the spyglass to her eye and once again checked the scene across the harbor as the ships became silhouettes in the glare of the setting sun. One double-masted ketch named the *Falconer* rocked slightly, anchored a bit apart from the *Love,* the *Expectation,* and several other little ships and yachts. In the East River roadstead she saw pinks that had come in with sugar and molasses from Barbados, sloops loaded down with fish from Boston, galliots from Curaçao carrying rum and salt and probably some exotic souvenir such as a pineapple or parrot tucked away in the hold. Each year ten to fifteen vessels entered the port with cargoes worth more than fifty thousand pounds apiece. Hers would be in that category. But there was no sign of the flute-ship that bore her name.

This was 1682. The *Margaret* had been anticipated in port yesterday. The skipper of a fishing craft had spotted her ship south of the Narrows, off Coney Island. The vessel had been under way now a month, and berthed in the West Indies before that a month, while the *Margaret's* supercargo assembled a load of Yucatán logwood to haul back to New York and reship to Dover. Logwood had been lucrative ever since the textile manufacturers of England found that its heartwood could be used to color their silks a variety of rich and fade-proof shades: dusky purple, or jet, or a blue to rival the blue of the Old World dye plant called woad. Since the fisherman had spotted the ship, Margaret knew it had at least arrived intact, but whether it carried its intended cargo remained a very significant concern.

Margaret's success meant she could stockpile silver spoons, as many as she desired, not only to conserve for funeral souvenirs but for everyday use. Her kitchen could have been a real Dutch kitchen in Amsterdam, as she could in fact now load her shelves with any item she cared to ship over from the fatherland. On her own ships, for that matter. And alongside those personal milestones—and with the hard work of an artist or a scientist, a person for whom hard work is at bottom nothing but exhilarating play—Margaret had managed against all odds to adhere to the distinctive values of Patria concerning women, which held that even a housewife was entitled to go out into the public domain as her own person, inherit fairly if she were widowed, buy and sell, spell and cipher, and continue (at least some of the time) to sign her surname with pride. No matter what the English might say.

Now Eva, at twenty-two the age Margaret had been when she stepped onto the Manhattan shoreline for the first time, helped her mother scan the horizon. Beneath them, treetops floated in a green cloud around the house's roofline. Eva scrutinized the harbor with the same intensity of her mother. Frederick had departed earlier in the day for Albany, to tend to business the couple had upriver.

There it was! Eva caught hold of Margaret's sleeve and dragged her to the railing. Spyglass to eye, Margaret saw the masts first, smoke gray, rising up on the horizon against the clear globe of sky. The ship was home, and God willing her cargo—the crew as well as the logwood—was safe. The *Margaret* was just one of so many ships that had returned to her or on which she had traveled home to Manhattan. Yet she always felt the same satisfaction.

With every year of English rule, women in trade were fewer, their liberties more circumscribed. In 1682 Margaret still actively governed a fleet, but she could no longer represent herself in a court of law. Manhattan, twenty-two square miles of opportunity, had been the freest place for women since Europeans set foot on the continent of North America. But those days were past.

From this leafy vantage, Margaret could see all the bodies of water that stretched around the island—the North River, the East River, New York Harbor. It was still an incredibly fertile place for her endeavors, furnishing warehouses, ships, and the rest of her needed trade apparatus. But now she saw the vast lands stretch out around Manhattan like the verdant rays of a star, while the island itself appeared as small as a beetle at the center of it all, as insignificant as it was frenetic. New York, the center, seemed to be closing in on a woman who was most emphatically unwilling to give up her she-merchant status. To keep her power, perhaps she needed a different center.

8

The House Margaret Built

$\mathcal{M}$argaret launched an epic buying spree in 1670, when she and a partner bought up three hundred acres of prime Westchester County land thick with forests of virgin timber and interspersed with fields the Lenapes had cleared for farming. Called Colen Donck—Donck's Colony, after Adriaen van der Donck, the first European to own it—the property's most productive feature was a sawmill standing astride the Nepperhan River, just above the Hudson. The mill's heavy blades drew on the power of the Nepperhan's rapids to cut boards from multiton oaks the way a serrated knife goes through soft bread. Planks tumbled in a chaos of slices out the front of the building, to be floated down the river to the Hudson and then hauled laboriously onto ships that would sail the commodity twenty miles south to the Europe-bound vessels berthed at Manhattan. Given the value of American lumber in the seventeenth century, the mill probably was worth as much as the entire parcel of wild lands that surrounded it.

That was just the beginning. In the decades that followed, Margaret and Frederick bought up neighboring tracts large and small, including the Upper Mills, the name they would give the remaining 7,708 acres of the Van der Donck property. This monster tract of land hugged the Pocantico, a powerful tributary that poured into the Hudson, offering both waterpower and a perfect spot for loading watercraft. The Upper Mills would be the future center of the family's wheat-flour empire.

Since the advent of English law, Margaret could no longer legitimately purchase land under her own name. But that legal restriction was no bar to the Hardenbroeck-Philipse juggernaut. They bought some parcels from investors and settlers, and many others from Indians. It was customary to apply to the colonial government for permission to buy native properties, after which a buyer made a deal directly with the sachems of the tribe. On June 5, 1684, Frederick Philipse signed a deed for a substantial parcel of land between the Nepperhan and the Bronx River, one purchased from the land's "native proprietors."

This particular document holds special significance as the only contract from a Philipse property purchase to survive in its original form; it was published verbatim in the mid-nineteenth century in a local county history. The details of the deal reflect increased sophistication on the part of native land sellers in the years since Minuit swapped Manhattan for sixty guilders' worth of goods. The remuneration still could be construed as a pittance compared to the actual value of the land, but by the mid-1680s an offer had to include a decent array of desirable features to be seriously considered by any sachems working the deal. In this case, imported textiles and clothing made up a good part of the trade—fourteen fathoms of duffels, twelve blankets, six kettles, six fathoms of the cloth called strouds, sixteen shirts, and twelve pair of stockings—in addition to the big-ticket items of twelve guns, fifteen pounds of powder, twenty bars of lead, fifteen hatchets, ten hoes, ten earthen jugs, ten iron pots, and one hundred thirty fathoms of white wampum. This was a huge volume of wampum, which if mounded in a hill would measure about 780 feet in length and 6 feet deep. The Indians still accepted kettles, six in this deal, but more coveted still were two ankers of rum, four and a half vats of beer, and two rolls of Virginia tobacco. With their returning ships' packed holds and their hogsheads of wampum, Margaret and Frederick were perfectly positioned to make this deal and the others required to stitch together the quilt of their grand domain.

At the time of Margaret's foray into Westchester real estate, there was great apprehension in New York about the Indians who remained in the vicinity. In the late 1670s Governor Andros had ordered surveillance of the Weckquaesgecks, whose traditional capital stood square in the center of the lands the Hardenbroeck-Philipses were determined to acquire. In 1675 a powerful sachem known as King Philip had orchestrated massacres in Maine, Massachusetts, Connecticut, and Rhode Island. New Yorkers feared they were next. Frederick Philipse, however, did not appear cowed by these events. A November 1679 letter from one Frenchman to another referred to the anxiety in the environs of Albany, "for they have sent a certain Mr. Philippes to examine the roads leading towards them. He had two Savages for guides."

Andros's council had already collected some Weckquaesgeck representatives to grill them "as to their Intention to join King Philip." The minutes of their meeting captured a touching loyalty on the part of a people about to be forced permanently off their ancestral lands. The Indians promised not to engage in any action against the Europeans, and also vowed that "when they certainly know of any disturbance or like to bee, they will give notice to ye Go. & they hope to have notice from thence of any hurt intended against them." The natives then made a special request of newly appointed council member Frederick Philipse, asking whether the unique bounty of the Hudson shoreline in the vicinity of

Westchester might be grandfathered to their people now that a colonist owned its deed. "They desire as before from Mr. Philips to have leave to come [. . .] here about Oystering."

In this charged atmosphere, no matter how assiduously some colonists might adhere to deeds and contracts, there were bound to be miscommunications or outright lying, a problem compounded by the fuzzy system of landmarks. After all, who could be sure exactly which big tree by the row of rocks on the hill near the river was meant in a given contract? There would be recriminations, years later, from some natives who were sure they had been swindled. The Philipse-Hardenbroeck campaign to acquire the lands along the east bank of the lower Hudson was all the more amazing for the lack of rancor and the permanency of the deeds.

By 1685, as forty-eight-year-old Margaret smoked her Gouda pipe and surveyed her domain, waiting for the advent of spring and the start of the trading season in the Hudson Valley, tracts in Mt. Pleasant and Ossining had also come into the couple's grasp, filling out their possession of all lands that stretched from the center of present-day Yonkers to the Croton River and from the Hudson River on the west to the Bronx River on the east. In 1687, as a grace note to their purchases on the east side of the river, Frederick and Margaret would venture west to scoop up a piece of the Tappan salt meadows in Rockland County, a stretch of waterfront from which they could wave across the river to their formerly Weckquaesgeck land. And by January of 1694, Frederick would look south to acquire the property that completed the Philipse domain, acreage that included the fifty acres of fertile, level Indian maize lands that reached from the Nepperhan as far south as Spuyten Duyvil. With this stroke, Frederick seized the literal keys to the highway, as the family now controlled the crossing of every person who entered or departed Manhattan.

By the time Margaret and Frederick finally stopped purchasing land and bought out all of their original real estate partners, the couple had amassed countless acres of choice fields, forests, streams, and bluffs. Literally countless, as time has refused to provide a definitive estimate of the total acreage of the estate. The namesake son of Frederick Philipse III, Margaret's great-great-grandson, would assess the dimensions of the family domain as "twenty-four miles long and upon an average six miles broad, which contains 92,160 acres." Later scholars, though, have detected a hint of grandiosity in his estimate, with one 1939 effort to tote up all the parcels of manor land dispersed after the Revolution figuring the total as just under 50,000 acres. Probably the most reliable computation, though, comes from cartographers at Westchester County's department of Geographic Information Systems Scholars, who have concluded that Margaret and Frederick held somewhere close to 57,000 acres. Whatever the precise measure, it no doubt formed a nice backyard for the modest 300-acre tract on the Nepperhan that Margaret had snapped up when the price was right.

*F*or a dozen years, the she-merchant did little with her new land. She was busy—building her trade, expanding her fleet, raising her children, and caring for her home on Manhattan. She traveled to the Colen Donck property by sloop sometimes, but more often than not she simply admired its tree-covered hills and crashing waterfalls as she and Frederick cruised past on their way to trade skins in Albany.

When she finally decided to move, she chose to build a house on the first piece of property she had bought, its location a sentimental gesture balanced by hardheaded practical concerns. She wanted a highly visible site, one that passersby on the Hudson could admire. She also wanted a reasonable proximity to Manhattan's slips, shipyards, and warehouses. The locale gave in addition a measure of autonomy. Here she could bypass the Manhattan-based British authorities with goods she preferred to bring in or ship out

undetected, untaxed. This would be a serious commercial investment, as well as a beautiful prospect for a residence.

And so in 1682, the flute-ship named the *Margaret* sailed from
Amsterdam with a cargo of building materials, including sixty thousand bricks, hand tools, and saws, as well as four millstones. After
crossing from Europe, the ship entered the East River roadstead, re-
distributing a good part of its cargo to a smaller ship to continue up
the Hudson. The smaller ship anchored at the obvious place to site
a house when you had the world to choose from: a bluff that rolled
down to a pairing of two exceptionally fertile rivers.

Beside the salt marsh below the bluff, at the meeting of the
smaller Nepperhan and the mammoth Hudson, curved a clayey
strand where Margaret's yachts, sloops, and flutes could easily load
and off-load. A wide sand path ran from the shore uphill to the
homestead, passing a clearing in the reeds that was all that remained
of an age-old village that had belonged to a local band of River Indians, the Dutch appellation for any native of the lower Hudson
Valley. The clan had disappeared from this particular gathering spot
long before Margaret started work on her house. But a mountain
of discarded oyster shells rose up beside the shore, a chalky reminder
of native feasts.

The house that Margaret built was as much a fur transit center as a home, a simple structure with a full cellar down a rough-
hewn wooden ladder. This was no root cellar, but a space carefully
fitted out for storage, laid with green glazed tile, which was the
customary flooring used for work basements in the better houses
of Holland and particularly in Frederick's birth province of Friesland. A bulkhead facilitated the transfer of valuable merchandise in
and out of the basement, and small windows just above ground-
level let in precious light. Here, Margaret could conveniently mon-
itor the cloth, guns, cord-wrapped bundles of pelts, and hogsheads
of wampum that were her stock in trade.

The house was not only a workplace but a showpiece: Constructed of coursed rubblestone masonry with rock harvested from the surrounding lands, the dwelling embodied the pride and pretentions of its increasingly successful owner. Out here in the wilderness, with only a weathered gristmill for company, Margaret's house stood two stories tall—albeit one room over one room. It boasted half a dozen extra-large casement windows squared up with the durable brick she had transported from overseas. (With amply proportioned, rock-solid window seats—just what a ship merchant would want to come home to after an arduous trading voyage.) For a New York country house, this one was incomparably grand. Its color was like the rest of the structure, simple yet strikingly handsome. From a distance it appeared gray. Close up, the gray had a hundred colors, because its granite came from the soil of the Hudson Valley, and sparkled with every tone from midnight to pewter to dove.

One feature of the dwelling suggested the more outsized, flamboyant aspects of Margaret's life: Its front door was ponderous, swung from iron hinges, and featured a giant lock, six by ten inches in size. (Legend contends it was carved of ebony and commissioned from Africa by the she-merchant herself.) A fail-safe lock to protect a cottage in the remote woods? The cellar was a veritable bank vault when furs and shell beads were money. If Dutch housewives kept peppercorns under lock and key, a fortress to protect pricey trade goods hardly seemed like overkill.

The throat of the chimney that faced the house's entrance was constructed of bricks held together by plaster fortified with horsehair, and it was huge, designed in the style of most Dutch fireplaces—so big a person could walk inside and look up to see the clear cobalt heavens through the brick-framed top. A flagstone hearth extended into the room, providing a generous space to prepare meals. Despite the heat that escaped up the chimney, with plenty of furs the house's occupants would stay comfortable even in the coldest weather.

The walls had no paint or wallpaper or other embellishment besides whitewash. The fireplace, though, was framed by painted tiles that must have struck Margaret as especially chic, with their exotic pictures in the stylish minor-key tint called manganese that resembled the blue-brown of a fading bruise. (The shades called "sad" likewise persisted as high fashion in the clothes of this period, denoting not necessarily gray or black but muddier earth tones, whether russet, plummy red, or the golden brown called "tawney." Even some pewter plates received the descriptor "sad-colored.") More staid Dutch homeowners who employed tilework in their homes went with either pure white or sober biblical allegories in safe shades of blue. Margaret chose a different theme. The painted scenes showed a Delft craftsman's cracked vision of the *wilden* of the New World: heavy-lidded hermaphrodites that frolicked on animal feet, breasts bulging, carrying fruits that resembled ripe melons and accompanied by old-style griffins. These images reflected the era, which paired intensive high-seas exploration and scientific curiosity with tenacious ancient beliefs in monsters. Artists and writers without firsthand knowledge of lands abroad still portrayed the landscape of America as crowded with Cyclopes and unicorns. The fanciful renderings on Margaret's hearth tiles offered an ironic counterpoint to the house's site, centered among the ghosts of ancient native villages whose all-too-human inhabitants had perished of fevers, plagues, and violence.

For the floors of her house, Margaret had her pick of the mature yellow pine from the virgin forests that touched the sky over the roof, and a pair of experienced draft horses to drag the logs through the woods. With uneven planks that ranged from eight inches to a foot in width, the couple accomplished their goal of using every scrap of wood they had harvested.

Frugal, practical, practically free of ornamentation—this was the house that Margaret built, with her husband's help. In many ways the house resembled its creator. It was crude, vital, improvisational, strong, and built to last. When Margaret erected her house, only a minute population of colonists resided anywhere nearby. There was a miller and a yeoman farmer or two. Rail transportation would not arrive for two hundred more years. No roads existed. Narrow trails crisscrossed the undeveloped woods, but these were intended for swift, narrow-framed humans rather than heavy carts pulled by teams of farm animals.

The year after Margaret's house rose on the north bank of the Nepperhan, the political organization of the surrounding landscape changed significantly. As of 1683, the colony issued a "Charter of Libertyes and Privileges" under the authority of Colonel Thomas Dongan, an Irish Catholic war veteran who was the province's reigning governor. Henceforth, the colony would be divided into a dozen counties, called "shires" in proper English locution. They included: New York (Manhattan), Kings (Brooklyn), Queens (the English towns of Long Island), Richmond (Staten Island), Suffolk (eastern Long Island), and Westchester among others, each with its own governor-appointed justice of the peace, high sheriff, and militia. For the first time in its history, the colony would have a representative assembly, with eighteen delegates named through county elections. The charter further mandated a governor and governor's council and English-style trial by jury. The governor would appoint New York's mayor, who would preside over the city council as well as the sheriff, clerk, and coroner. These new political entities would absorb Frederick and the other men of the Philipse family, who would fill assembly seats and other powerful positions. The land Margaret and Frederick bought north of the city was officially part of the county of Westchester. Margaret's house stood in the village of Yonkers, the first municipality north of Manhattan, its name a relic of early New Netherland when Colen Donck was owned by the young sir, or Jonkheer. But the couple's vast estate also formed an entity all its own. This was a landscape where one family would enjoy a superfluity of power and influence for generations to come.

*A*nd so on an early spring afternoon in the mid-1680s, Margaret would wait on her stoop overlooking the Nepperhan for her sloop to arrive, the ship that would carry her to Albany for the start of the fur-trading season. She wore layers of clothing and kept a hot fire going in the house, but the chill of the air outside caused her to pull her cloak more tightly around her. A still largely wild, un-plumbed, endlessly leafy vista fell away from the bluff where she sat sentry, and it was easy to feel that she was the only person in the whole green universe.

Soon, she knew, the shad would be on the move, swimming north to mate. The fish announced the season to anyone who lived on the river. At the first budding of the forsythia, the bodies of the shad would be barely visible, a seething just beneath the surface. Later, with the flowering of the cherry blossoms, came a second wave of fish. Finally the lilacs bloomed, and the fish threw them-selves upstream so powerfully that the humps of their platinum backs gleamed in the river's platinum tide. The River Indians called the shad a porcupine turned inside out. Its flesh tasted so sweet roasted on a hickory plank over a smoldering wood fire it was easy for colonists to ignore the fish's dozens of pinlike bones.

Margaret had no way of knowing definitively whether the pelts she had been promised when she traveled to Albany last fall would materialize when the trading season opened in a few weeks. She also did not know exactly when her Indian trading partners would arrive in their canoes loaded with pelts from settlements much farther west in the wilderness. In late fall, in the woods outside Albany, an interpreter had assured her the deal would proceed as promised, that all would be overseen by Minewawa, the goddess of the valley who hangs up the new moon and takes down the old to cut into little stars. The hunters were ready to move, to disappear into the frigid, snow-mounded woods. They would meet with her again as soon as the ice melted on the water.

Margaret had worked with some of these native men for decades, since she started trading skins in the mid-1660s. She trusted their word. Now that the ice had almost disappeared, she brought her currency from Manhattan—knives and wampum, brass kettles and bolts of the cloth the Indians requested. These were extra goods, to sweeten deals and pick up additional merchandise. Most Indians demanded payment in advance, so she had already paid in the fall for the furs she would claim in spring. It was a risky way of doing business, but there was really no other way if you wanted a relationship with the most skilled hunters—it was a seller's market. She even brought rum, termed by the Indians "English milk." Some hunters would accept nothing else, although it was known to make a native man drink until he consumed the last drop and then go crazy, giving away his best furs for next to nothing.

The shipment Margaret awaited would consist mainly of prime winter beaver, taken when the young animals were in heavy coat. The trappers' wives would have already cured the skins, scraping the raw side until the roots of the long hair dropped out, leaving the short barbed hair that was ideal for felting. The best felt came from greasy, battered skins, and the most artful tanners pounded the pelts with animal fat and brains and liver to soften them. The hunters sometimes sewed the skins together to make robes, wearing them fur side in until this "coat beaver" grew downy. Since the early days when Indians tore off their own cloaks to make deals with traders, no pelt was more valuable on the European luxury fur market than preworn beaver skins seasoned with the sweat and body oils of the American *wilden*.

*I*n the winter of 1680, a great comet had roared across the sky over Albany, a tiny settlement 150 miles north of New York, perched at the curve of the Hudson as it veered toward the western frontier. The town had about seven hundred year-round residents and a prominent yet all-but-useless old fort sunk permanently to its ankles in a great bed of river mud.

The comet's glowing cat's tail had captivated colonists in North America's fur-trading capital. They hungered for celestial explanations for daunting events in their lives and attempted to draw off God's ire with bouts of communal thanks to God and repentance for sins committed, hoping to restore order after the chaos of flood, drought, disease, or famine. A common theory, for example, held that one terrifying orb—later identified as Halley's Comet—that lit up the sky over England in 1066 had foretold the Norman conquest of England. The meteor of 1663 had likewise presaged drought, flood, and the Indian massacre at Wiltwyck.

When the comet of 1680 illuminated the streets and fields of Albany night after night, week after week, the commissaries of the town wasted no time. They sent a letter by Indian post to Captain Anthony Brockholst, lieutenant governor of New York as well as commander in chief of the provincial military, despairing of the "dreadfull punishments if wee doe not Repent." They pleaded with Brockholst to declare a day of fasting and humiliation, preferably on a monthly basis. In 1680, a bad year for the colony of New York, crops had failed, corn was scarce, and the people of the town could only imagine what dire events might follow.

Oddly enough, it seemed a streaking star could signify good fortune as well as bad. For Margaret and Frederick, the comet of 1680 seemed to herald an ongoing immunity to bad fortune that dated back to 1665, when Governor Andros awarded the couple a license to engage in the upstate fur trade with no middleman. By the 1680s, their commerce had reached a new level. Their fleet was now reliably bringing back massive cargoes of goods that they would sell (the ells of cloth, frying pans, and bridle bits ad infinitum on the flute-ship *Charles*), gaining the capital they needed to invest in the wampum that let them purchase skins, along with the duffels and guns that were an integral part of every trade. They were golden—or, to put it more appropriately given the furry commodity on which they built their business, plush. Margaret and Frederick had situated themselves in exactly the right place. As one late-nineteenth-century colonial historian put it, "In the early days every house in Albany was a trading store, with furs in the second story." The club of high-volume merchants in Albany—central among its members Frederick Philipse and Margaret Hardenbroeck—dispensed with the second story and set up commanding freestanding pack houses where their loads of hides were readied for shipment to points south, either New York City or, in Margaret's case, the house on the Nepperhan with its fur-ready storage cellar.

Margaret was at the top of her game. The number of female traders had dropped in recent years, but this particular she-merchant would not see her name exit the roster. In large measure this was because she acted in concert with her husband, who served as a sort of economic beard, diverting attention from her activities with his own financial exploits. But it was also a matter of strategy: Margaret had shifted her enterprises to Albany, Westchester, Amsterdam, the smaller port towns of England, and the West Indies, making her success on the margins of the British culture that had removed her from contracts and courts.

Margaret especially enjoyed Albany. No American town was more Dutch in spirit than Albany, certainly not Manhattan after more than a decade as a royal colony. Margaret traveled frequently to Amsterdam, but when she was in Albany it was almost as if she had never left Holland. She and Frederick had leased a place there before, most notably in 1765, but the increasing volume of their business warranted a permanent dwelling. Plus, the town now prohibited merchants from participating in *Handelstijd,* Albany's spring trading season, if they did not reside there, at least officially. In May 1676 Margaret and Frederick bought a fine house on Yonkheer Street, within view of the Dutch Reformed Church. To further develop their Albany enterprise, the couple then invested in an eighty-foot-wide lot at the edge of town, along the river. The locals had long been resentful of outside merchants, and an Albany address had become a requirement for doing business in town.

Even the street names in Albany spoke of the town's one primal purpose, the exchange of furs. Handlaer, the main concourse that ran parallel to the Hudson, meant "street of trade." Yonkheer, "gentleman's street," had become an avenue of grand homes, a consequence of decades of the booming trade in skins. Many of these homes served the dual function of offices for the *handlaers,* the traders, to conduct their business. Spring was the time for the bigger merchants to buy pelts in Albany. It also was the season to sell for the traders of lesser means, who just as fervently desired to take part. A man might sell the bed he slept in, or the boat that allowed him to fish for his dinner, or even an old pair of trousers. Handelstijd meant people would sell—something, anything worth buying—so that they could secure the funds to buy one beaver, even a small fur, or a fur in less-than-perfect condition. That tattered skin might be worth something to one of the wholesalers heading downriver with a boatload of merchandise. A merchant such as Margaret might be persuaded to lay out a handful of wampum to buy four muskrat skins from an Albany middleman, who had bartered his own worn shoes to a native trapper to get them. The wampum could put bread on his table for at least a month. One single beaver pelt in "good merchantable" condition, not thread-bare or torn, might serve as payment for a number of a Dutch household's necessities: a quantity of tobacco, thread, or bacon.

Starting each May, the town traditionally doubled its population. Handelstijd attracted so many visitors that the market inevitably overflowed its central location. The British passed ordinances early in the 1670s to fix the location of trade activities, but it proved next to impossible to calm the atmosphere, even in later years when the market in furs gradually slowed. During the warm months, the riot of trade reached every corner of every street, touched every house-hold, and penetrated the deep forest that lay outside the edges of the walled town, where natives lugged their pelts to market on well-worn paths.

From early April until the end of June, Indians came in from the west and north, camping in battered wooden shelters built for their use—considerately, in the familiar shape of Iroquois long-houses—on a hillside outside the city stockade. It was an ongoing scandal that certain Dutch traders went into the woods and over-whelmed the trappers, plying them with booze to get better deals on merchandise. Fur traders sold Indian trappers eight thousand gallons of rum in a single year, 1687, according to one contempo-rary estimate. Some settlers invited Indians to bunk in their homes, where they hoped their rum and hospitality might persuade their guests to part with skins on the cheap.

English and Dutch traders sailed up the Hudson from Manhat-tan just as the spring forest unfurled its blossoms—wild plum everywhere, the snowy spangles of dogwood, and the tulip tree, with its leaves that could be crushed as a headache remedy and bark for treating horses with worms. Soon, strawberries would be on every table, ready to gobble up with wine or sugar. The air was filled with smoke as the Indians torched the woods to thin the trees, reduce the whisper of grass under foot when they hunted, and trap game within the lines of the fires. Sometimes the blaze climbed the trunks and set the crowns of the trees on fire, and people passing on the river could see wind-driven flames dancing high over each bank.

Margaret and Frederick spent a good deal of their time in Albany during the spring season. She hired her father, Adolph, to oversee their house during the rest of the year, when the town's population dwindled to the hundreds rather than the thousands present in the season. Adolph and Maria must have felt right at home. Since Albany's founding, the center of the town had been known as the Fuyck, the Dutch term for a trap set to capture deer or fish, because of its evocative wedge-shaped outlines. Handlaer and Yonkheer streets came together close to the river in the configuration of a wishbone; each had houses and shops of red *moppen* and yellow Gouda brick crowded close together, all with eaves that extended so far into the street they dumped water on passersby during rainstorms.

The Dutch Reformed Church, built in 1656, was constructed of logs, but it showcased an elegantly carved pulpit imported from the Netherlands, and a stained-glass coat of arms in every window bearing the names of each principal contributor. Near the top of the hill, the English had built their new fort—an attempt at establishing a presence in this Dutch town. But the houses all around the neighborhood of the fort showed who really counted, still. If you lived up here in the nosebleed heights of affluence, as did the Philipse family, you could roll your hogsheads of pelts right down to the waterfront if you felt like it.

In Albany, the children strapped on skates and took to the river when it froze deep enough, as they did in New York, but this far north, where the ice could be three feet deep, the recreation lasted months. Teams of horses were known to cross the river ice as late as April. "Coasting" was a sport the boys and girls of Albany perfected. In large, rowdy parties they took to the hills on sleds of rope and wood, sliding down among the shaggy boughs of white pine with its sharp, cold scent. The children came home to buttered bread and the new sensation, a beverage called chocolate made with sugar and hot milk.

On freezing Albany nights, families burrowed down in alcove beds made toasty by brass bed warmers filled with hot coals. Travelers reported fires roaring on every hearth, even in the hot season. In summer, fireflies flew in the open windows, along with clouds of biting gnats known as punchins against which residents smeared their faces with butter.

The houses had the same Dutch flavor as the houses of New York, and the town had grown more prosperous as the fur trade expanded in recent years. Silver became more popular as Margaret and her peers grew more flush. Heavy embossed teakettles appeared on tables, along with silver muffineers for sprinkling salt on buttered muffins. The well-to-do often added a separate kitchen, moving their cooking facilities out of the common *kamer* into the cellar. The woman who ruled the premises kept her spice box hidden, hoarding the pricey mace and nutmeg that flavored the most popular dishes. A marble mortar held pride of place on the long plank table, the table where most food preparation took place, its pestle ready to crush medicinal herbs.

Tavern life was still rowdy after the British takeover, but domestic manners aspired to a new civility. Even in Albany, households had begun to serve tea in hard-paste porcelain either brought over from China or produced by English imitators (these smeared-looking blue and white tea dishes tended to be imperfectly fired, especially when the process was new). They had no handles, so colonists drank the beverage on the cool side. Rich and poor women alike found that cider, distilled as applejack, made the perfect nectar in which to bathe peaches and plums and cherries. They learned to create rose water—used for cooking as well as for cologne—in special stills of brass, and they put up raspberry liqueur in season.

Margaret could fit in to the comfortable domestic life of Albany, if she felt inclined. When she and Frederick bought the house on Handlaer Street, their children were teenagers. The Philipse kids knew the town's secret spots. They listened to the braying debates in the streets outside the taverns, and learned which hausfraus could be counted on for a cookie or a cup of milk. They learned, too, when they accompanied their mother as she toted her bulging account book to the Strand, to politely greet the people who came to them with mounds of musky pelts, chestnut, silver, red, and the deep licorice of bear.

As the years elapsed, Margaret's sons joined in the family business. Adolph had already begun to make his way in international trade, dutifully commuting between Europe and New York to help to manage the expanding Philipse shipping concern. Dutch court documents of 1688 would describe the twenty-year-old as a "free merchant" of Amsterdam, engaged in establishing prices on a shipload of furs.

Philip seems to have followed even more doggedly in his mother's and father's footsteps than his budding merchant brother. He helped enlarge the scope of the family's landed estate in 1686 when he paid the Lenapes for a tract adjoining his father and mother's land (he was just twenty-three at the time). Later, not content to serve as factor, he trained to take a more hands-on role as captain of his own transatlantic bark, the *St. Mary.* Customs records show that Philip Philipse cleared Plymouth, England, as captain of the *Land of Promise,* toting a cargo of 1,311 hogsheads, 1,011 tierces, and 13 barrels of the best Barbadian brown sugar.

Like many of the ambitious young men of his generation, Philip indulged in privateering when given the chance, as in July 1690 when the governor of New York commissioned him and two other men to "cruise along the coast" of Long Island "and capture several French vessels about Block Island and the Sound." (No matter what the outcome of this adventure, it is hard to think that Margaret was not made proud.) Philip's passion for the seafaring life was no doubt fueled by early experiences he had had traveling in the company of his factor mother—like his sister Annetje, on the *Charles*—and by the understanding that as the eldest son in English society he would be heir to the family shipping empire. Like the other children of New York, he had the benefit of a primary education, but more pertinent was the maritime schoolroom in which his family specialized.

Albany would remain a hub for Margaret's enterprises. But her social life had grown more focused and more selective as her family's wealth assumed proportions only the equally privileged could imagine. Margaret and Frederick spent their time with their peers in the business world, the cream of New York society, such as Jeremias van Rensselaer, scion of the great landowner who early in the life of New Netherland founded the estate called Rensselaerwijk. Van Rensselaer feted guests with rare brandies at Watervliet, the manor house he built after the Hudson swept his first home away in the spring flood of 1666.

Margaret had known Jeremias's wife Maria for years on Stone Street, where she grew up a daughter of Oloff van Cortlandt, but never saw the depths of her character until after the birth of Maria's first child, in 1663, when an infection known as septic arthritis erupted in her hip, causing the woman crippling pain for the rest of her life. By the time Margaret bought the Albany house, Jeremias had died, Maria had six more children and she was running the million-acre estate single-handedly. She even managed a brewery she had begged her husband to build for her.

Maria, hobbling around the rough Rensellaerwijck acreage using a wooden crutch under one arm and a cane in the other hand, could never be a companion to Margaret as she swaggered down Handlaer Street in the Fuyck. But the two shared the passion for trade, physically divvying up shipboard space on either Hardenbroeck-Philipse or Van Rensselaer or Van Cortlandt vessels to get their goods from Albany down the river. Each family knew they could trust the shipmaster and sailors hired by the other and the quality of the vessel. The men of the Van Rensselaer family, and the Philipse family, and the Beekmans and the Livingstons, served on councils together and advised the governor when called upon. The women no doubt advised the men.

9

A Surfeit of Sugar

*E*verything was sweeter since the Hardenbroeck-Philipses bought the estate on Barbados.

There was cone upon cone of the finest-grade sugar, sparkling pure white, always available to scrape over porridge, to sweeten creams and custards, to stir in tea. And there was the cane itself— when the workers cut down the jungle of tasseled stalks at harvest you could break off a piece and suck the juice. Even adults did it, when they weren't drinking Kill-Devil, the white rum brewed from molasses, the amber by-product of the sugar-refining process.

The air on Barbados itself smelled sweet, like cotton candy or caramel. All over the island between October and December, workers planted fields in staggered rows with short sections of cane buried in perfectly regular furrows, like long graves. White plumes of flowers filled the air with scent the following December, and between January and May, sixteen months later, the plants were ready for harvest. By then the forests of stalks had ripened from the color of new grass to a shade seventeenth-century Barbados diarist Richard Ligon called "deep Popinjay," a fluorescent parrot green.

A network of sugarhouses processed the cane and fixed the smell of sugar in the air. The crop was highly perishable. Once cut and carried in bundles from the field, the cane had to be crushed by sundown in a mill. Then the juice produced had to reach the boiling pots quickly or it would ferment—and all the grower would have was a ruined crop.

By the time Margaret and Frederick bought their estate, Spring Head, in 1674, virtually all the arable land on Barbados had been given over to sugarcane. For them, the island was the next big thing, the next step on the ladder to new wealth. New York had given the couple a foundation. Acquisition of the Hudson Valley property increased their net worth, although agricultural production on their 57,000 acres had not yet reached its potential. Albany provided the raw materials that formed the basis for their business. Now they would branch out, a good idea since the fur market was destined to take a downturn. Even in 1674 it was clear that the animals were beginning to vanish, that even the most adept hunters could not continue to supply pelts in the quantity merchants needed if fur trading was their only means of staying afloat. Fortunately Margaret and Frederick could afford to diversify. This new opportunity, a British colony located more than two thousand miles from Manhattan Island, did not have convenience to recommend it. Yet in the third quarter of the seventeenth century, it seemed any merchant who could possibly summon the resources threw them at "the brightest jewel in His Majesty's crown."

The couple's new property on Barbados floated above the southern escarpment of the island, with a 180-degree view of aqua seas and a buffeting ocean breeze. No estimate of its acreage has survived, but a map of the island commissioned in 1675 by Charles II to assess economic opportunities there—cartographer Richard Forde created an amazing document teeming with symbols of windmills and mansion houses—places Spring Head in one of the island's least fertile areas. That location and the omission of the Philipse name from tax lists of the largest planters of the time suggests that the couple's estate was a smaller property, one intended to serve as a base for the family's merchant interests rather than produce huge crops of cane. Margaret and Frederick likely didn't want to dirty their hands with the mess and risk of cane farming when there were so many guilders to be made buying and selling—especially when shipping was their expertise.

The outlook of the house must have taken in plantation after popinjay plantation, each estate with its dense jungle of blond-tipped stalks punctuated by the clutter of sugar-making equipment that skirted the fields. These elaborate operations had grown nearly as ubiquitous as the fig trees with beard-shaped leaves that gave Barbados its name, The Bearded Ones, when Portuguese explorers stepped onto its shores in the 1500s. Everywhere Margaret traveled on the island, she was surrounded by the thatched-roof pavilions of the boiling-houses, each containing rows of copper cauldrons atop raging cook fires that baked the neat red brickwork of the floors. Thickening solutions of cane juice and water simmered in kettles— hot, hotter, hottest—until the liquid reached the strike point, the precise instant at which the syrup transformed itself into granules.

Workers ladled the liquid sugar from one copper to the next, lifting scum off the top (to go to the still for rum), then potting the liquid in gumdrop-shaped wooden molds stoppered with plantain leaves. High-grade white sugar was separated from muscovado, a less-expensive sweetener, moist and dense, with the flavor of raisins and the color and texture of gingerbread. Muscovado enjoyed a huge popularity throughout Europe, especially in households that could not afford the brittle cones of white sugar clothed in blue paper and stocked by apothecaries and other shopkeepers.

In the harvest season, the aroma of sugar syrup, muscovado, molasses, and white sugar blended together. The intoxicating smell that wafted up all over Barbados offered a sensual promise to the island's settlers that a new age of affluence was at hand. For Margaret it was without a doubt the smell of money.

The dazzling coral island of Barbados, flung one hundred miles to the east of the West Indies and technically beyond the fringe of the Caribbean, had been haunted by ambition and violence since the arrival of the early sixteenth-century Spanish conquistadors, who kidnapped the Indians to labor as sugar-plantation slaves on the island of Hispaniola. English colonials who claimed the island for King James in 1627 were pleased to find no trace left of an indigenous population. That left them free to settle in quickly without the nuisance of conquering natives. From the start His Majesty granted land to socially connected English families of a certain class; many privileged young men ventured to the West Indies to make their fortunes. They were joined by gamblers, kidnappers, and political refugees.

"This island," lamented visitor Henry Whistler in 1655, "is the dunghill whereon England doth cast forth its rubbish," populated only by "rogues and whores." Single, bored, and far from polite society, the gentry seemed to strive to set records for decadence and consumption. Diaries record them throwing down gallons of Madeira at stupefied epic feasts. At a memorable repast hosted by one of the island's chief planters, James Drax, a phalanx of slaves brought out fourteen different dishes of beef, eight of fowl, three of pork, and three each of goat, mutton, veal, plus a suckling pig, oysters, caviar, anchovies, and olives, all of it lubricated with copious amounts of alcohol.

The Englishmen who settled Barbados refused to go native. They tried to ignore the heat and preserve the dignified pallor of their skin, wearing leather gloves, starched and ruffled collars, and broad-brimmed hats over hot and itchy periwigs. They remained loyal to the breeches, stockings, waistcoats, and flowing blouses of their class. If they could afford it, they built faux Gothic castles with walls as thick as a man's arm. At night they locked their windows against the sea breezes they considered medically imprudent, and further cocooned themselves behind the drawn damask curtains of their four-poster beds. Sleeping, they were stifled but safe. By day, a gentleman could avoid burning his flesh even under the searing sun since he had servants imported from the British Isles to carry him about in litters.

Gradually the planters hacked down the native forests of bully, fustic, cedar, and mastic that had rambled down to the golden beaches, merging with the thick palmetto groves. But the giant trunks were so cumbersome to drag away that vegetable gardens still grew between the immovable rotted limbs many years later. By 1650 the planters had their wish: no useless forests. But that also meant there was no wood to fire the cauldrons. To create an alternative fuel, the planters dried out crushed cane stalks in special shacks and then recycled this "trash" as coals—an endless supply of sugar trash feeding an endless harvest of sugar. The Brazilians, whose genius had put the sugar-distilling system in motion, had a term for it: *bagaceira,* meaning, "a life built round cane waste." The plantations even used spent cane mixed with manure to fertilize the fields.

From the start, Barbadian planters borrowed Dutch capital to finance their enterprises. The English had colonized Barbados, but the United Provinces took the crucial next step toward sugar cultivation when Holland native Pieter Blower brought cane to Barbados—though in his time, the main concern was distilling rum from the juice for local consumption rather than refining it for export. Dutch merchants introduced the English to the refining supplies and provisions they needed to become full-service sugar lords—with an ulterior motive, of course. The Dutch were less interested in growing the cane themselves than in making money off the loans the planters needed to get into the sugar business.

By 1660 sugar profits made the island the richest spot in the English empire. Fully 65 percent of all sugar from the West Indies came from tiny Barbados, its dimensions only fourteen by twenty-one miles. Each acre planted with cane produced one full ton of sugar, an amazing yield. Fortune seekers raced to the island, which by the time Margaret arrived had a population that equaled that of Massachusetts or Virginia. All because her countrymen had supplied a plant, technically just a grass, and imported the technology to make it valuable.

When the English first colonized the island, they brought indentured servants from home, many of them Irish, but these men and women rose up against their masters or ran away to lose themselves in the largely impassable mountains. The first English settlers also brought ten Africans to the island as early as 1627 to perform the unforgiving labor of sugar cultivation—to bend over the growing cane, clear the jungle of weeds, attend to the mills, and decant one scalding kettle into the next. Since then, hundreds and then thousands more slaves had been pressed into service. Even thirty-two "Indians" from Guiana, freemen drafted expressly to teach the English tropical agriculture, eventually found themselves enslaved.

Although most Africans sold in Barbados by the boom years of 1640 to 1670 profited Dutch merchants—at the time Europe's biggest slave traders—the English had initiated the trade in the West Indies a century earlier. William Hawkins of Devonshire probably was the first Englishman to capture and enslave Africans. In 1532 he bartered one of his own sailors for a chieftain he brought from the Guinea coast. The African died en route to England, while the seaman somehow made it home alive. When the English trade began in earnest, Queen Elizabeth I expressed her enthusiasm (she delicately hoped the slaves would be brought to England willingly) and sent her private ship to Sierra Leone on a voyage that included burning and pillaging native villages to obtain chattels that were then exchanged for precious metals and jewels. Slavers gradually realized that the biggest profit from Africans would derive from their value as laborers, once hardheaded colonists did the math: An enslaved human bought for twenty pounds (even a slave who was, according to New Amsterdam's first minister, "thievish, lazy and useless trash") represented an irrefutable deal, costed out over the average slave's lifetime.

The black workers of Barbados were forbidden to eat the white sugar they produced. Living conditions for the island's slaves were abysmal: no clothing, only minimal rags to protect them against the burning sun or blasting trash fires, and no shoes to shield their soles from the hot brick of the boiling-house. Rations were exceptionally poor, even compared with other plantation societies. Some planters gave their slaves rotting cow cadavers; others chose not to supply food at all, instead offering them rum to trade for food. While the landowners of Barbados made their way through fourteen-course debauches, their slaves were chasing land crabs in the moonlight.

In harvest season, processing went on around the clock. Uphill from the cauldrons stood sugar mills powered by pairs of oxen that plodded in a circle, their motion turning three metal rollers that smashed the cane between them to press out its juice. Weary workers who fed the long sticks of cane into the rollers lost fingers to the grinding mill; the custom was to keep a hatchet handy to keep from losing the arm as well to the motion of the rollers. All-night fires blazed across the landscape to light the boiling-houses, which also claimed digits, even limbs, and it was said you could recognize a sugarhouse worker by his maimed countenance. One planter was said to make his enslaved boiler test the sugar by dipping a thumb and forefinger into the cauldron to determine if the liquid spun a thread of the proper texture.

It was a truism that Barbados killed its slaves. Planters were aware of the toll because they had to keep replacing their workers, who usually died only a few years off the boat. Of the 170,000 Africans brought to Barbados in the half century after 1650, only 40,000 of them or their progeny were alive in 1700. On one of the grandest sugar plantations, six blacks died each year for every slave birth. On another the average life expectancy for a slave was seventeen years. Whites also died unnaturally young on Barbados, often brought down by drink and other excesses. Vicious storms struck with regularity. In 1675 the most deadly hurricane in memory flattened cane fields, wasted sugarhouses, churned up the principal harbor, and killed hundreds of whites and blacks alike. Fires raced through Bridgetown, gutting the genteel clock and goldsmith shops and splintering small wood-frame homes like fistfuls of matches. It was a deadly place to make a living.

$\mathcal{M}$argaret seized this new opportunity. First she supplied planters with Rhenish wine to accompany their gout-inducing dinners, bringing over dozens of casks in a "sweet ship," named for the aroma that lingered in the hold after transporting wine across the sea. (The *Mayflower* itself had been a sweet ship, having hauled its cargo of wine across the Mediterranean on previous voyages, and its alcohol-impregnated air must have inspired the pilgrims with visions of the land of Goshen they were sailing toward.) She also brought musty, golden, crumbling tobacco from Virginia. The men of the land fired up their pipes as voraciously as any she knew in New York, and she could provide as much of the leaf as they could buy. The aristocrats of Barbados needed to import lumber to build their mansions, because the forests had been eviscerated, and grain for bread because every square acre of soil went to cultivating sugar. They also needed beef—there was no room for livestock, aside from the animals that worked the roller mills. The English gentlemen preferred their familiar salted beef in barrels anyway, and had little taste for the fish that filled the ocean not a foot offshore. Some had converted to eating sea turtle—as long as it was dressed properly, the fins, belly meat, and head boiled in springwater, then stewed with a veal knuckle, cayenne pepper, and a pint of Madeira wine. Turtle had acquired the delicious flavor of status. But for the many who disdained the native foods of the island, Margaret had the beef.

In return for all of these goods, she acquired the dark red molasses welcomed by dozens of rum distillers in the northern American colonies. She could have the alcohol processed and then trade rum to the Indians at Albany. Manhattan merchants had been eagerly test-marketing Caribbean delicacies such as conch and manatee ever since Director Stuyvesant's wife received a "parrot from the Spanish coast" as a trendy gift. Margaret had customers panting for lemons and limes, if she could get them back to New York unspoiled. And to complement the tartness of lemons and limes in confections and punches you naturally needed sugar.

As sugar became cheaper and more readily available, as cane fields crept across Barbados and the other islands of the West Indies, and as Margaret and other merchants showed up with vessels to transport the product, the appetites of Europe grew predictable, easy to satisfy. The more people ate sugar, the more they craved it. Sugar was hardly new: For hundreds of years, the great chefs of Europe had assembled frosty concoctions of sugar paste for whimsical "subtleties" at wealthy tables—castles, battles, and biblical allegories that were both edifying and edible. Until now no one could afford the stuff but the aristocracy. But by the late seventeenth century the populace—especially in England—found its sweet tooth.

Part of the demand came from the new chic of tea drinking. Women sugared their Chinese tea as men had always sugared their wine, though the Chinese themselves had never thought to sweeten the ancient beverage. Europeans also adored chocolate, a beverage that was not only for children. It flowed originally from cacao plantations in South America to France by way of Louis XIV's Spanish bride, Princess Maria Theresa. She gave the Sun King a chest of chocolate in 1643 for an engagement present, and his avid consumption of the beverage was said to fuel his ability to pleasure his wife twice a day even into his seventies. Chocolate made its way from the French royal court to London, where one trendy chocolate house in 1657 spawned dozens of rival establishments, especially after the esteemed physician Sir Hans Sloane determined that substituting milk for water upgraded the creaminess factor of the delicacy. Soon almost everyone wanted to access the toothsome sweetness and the erogenous properties of chocolate. "Twill make Old women Young and Fresh," wrote New York philanthropist and bibliophile James Wadsworth in the nineteenth century. "Create New Motions of the Flesh, / And cause them long for you know what . . . / If they but taste of chocolate." And that was, of course, sweet chocolate, chocolate prepared unfailingly with cane sugar from the islands.

Margaret and Frederick could have continued to make a profit shipping molasses from Barbados to the Boston brewers and then bringing wine and beef to the island's planters on the return trips. Like the rest of the family properties, even the relative palace on Stone Street, the Barbados estate functioned as a business office nearly as much as a home. Margaret and Frederick divided their time between their various residences, alternating Albany, Yonkers, New York, and Barbados with occasional layovers in London and Amsterdam. Managing their vast commercial interests was easier now that the children had grown and sons Philip and Adolph took part in the family business.

The family's enterprises in the colonies and overseas continued to thrive. But like any successful entrepreneur, Margaret was always looking for some new product to improve her bottom line. And that might begin to explain why she decided to expand their enterprises in a dramatic new direction.

*I*n 1684 the *Charles* sailed from Manhattan to Amsterdam to exchange New York furs for textiles, casks, muskets, and, for the first time, shackles. After clearing English customs at the Isle of Wight, Margaret's flute-ship made its way to Angola, where 146 prisoners of tribal wars were escorted aboard the ship. Barbados was the optimal West Indian destination for ships carrying slaves, the closest landfall to the continent of Africa, thirty-five hundred miles across the ocean from Dakar, Senegal.

The *Charles* made just one stop, but slaving ships customarily cruised along Africa's Gold Coast to secure small numbers of prisoners from tribal leaders at many ports, guided by the common wisdom that Africans of diverse tribes would not be able to converse with each other to plot rebellions. In fact, that wisdom was flawed. Though eleven hundred languages thrived in the African interior, many dialects shared grammatical elements, and often the people sold into slavery were multilingual. Plus, prisoners desperate to communicate will find a way. On English slave ships, the number of mutinies—calculated by scholars at one every two years—was high enough to remain a serious threat.

Even the female captives, typically less heavily guarded than the men, wreaked havoc. On one "floating prison" that approached Barbados, women who had been brought on deck unchained for exercise grabbed muskets from a chest and overcame the crew. The mutineers' victory did not last: Unable to navigate back to Africa, they drifted for six weeks before being set upon by a British warship and resold into slavery.

On slaving voyages the captives—and many of the crew—suffered bouts of dysentery, smallpox, and other diseases en route, resulting in a mortality rate of about one-tenth of the people on board. The *Charles*'s maiden Africa run was no exception. Eighteen of the Angolans aboard died. Fourteen others were so ill they would not make it to market.

When the ship entered Carlisle Bay in Bridgetown, the island's thriving port city, Margaret would be there to watch her ship anchor in the roadstead, accompanied by her eldest son Philip, who was always eager to learn something about the business. Even from a distance, the olfactory evidence of human misery would sear an onlooker's nostrils.

Captain Thomas Codringham, the master of the *Charles,* successfully sold 105 of the surviving captives in Barbados. That left nine prisoners. And those nine would become the first enslaved Africans on Philipse-owned land. The colonial trade in slaves provided many benefits for merchants, who profited not only in the slaves they sold, but also in the slaves they put to work in their fields, mills, and town houses. When Margaret took up slave trading, she found it only natural to take up slaveholding.

In the province of New York, colonists had only gradually developed a taste for slave labor, partly due to scruples ingrained by the Dutch Reformed Church, and partly due to frugality. The West India Company, meanwhile, worked for a century to trump all other nations as slaver to the world—its directors seemingly undisturbed by the fact that Amsterdam and all other Dutch cities adamantly refused to allow slave ships to sell their cargo out of the nation's ports. The monetary rewards derived from slave-hungry West Indian planters outweighed any moral qualms Company investors might have felt about trafficking in human beings.

The investors welcomed the free labor to jump-start construction of the Manhattan settlement; neither freemen nor paid servants

had much enthusiasm for the scut work of raising Company buildings, clearing forest, and broadening Indian paths to accommodate
carts. In 1626 eleven African men became the first Company slaves
on Manhattan (their recorded names, Paulo D'Angola, Anthony
Portuguese, John Francisco, etc., suggest they were captured aboard
a hijacked Portuguese ship). Three Angolan women arrived in 1628
to serve as cooks and housekeepers. The colony's official slave quarters soon were established at dead-end Slyck Straet, a marshy neighborhood of hovels punctuated by empty lots (where one marginal
character, a homeless woman, slept alfresco winter or summer). The
Company erected a building to shelter its chattels, hiring a succes-
sion of colonists to serve as slave overseer.

Since then, the entry of slave ships into New York Harbor had
grown more common, and ship owners were expected to pay taxes
on their human cargo as they would any other merchandise coming into port. It was for that reason that the ever-cost-conscious
owners of the *Charles,* sailing from the West Indies with a cargo of
nine Angolans, instructed the skipper not to weigh anchor in the
East River. Instead, Captain Codringham sailed north of the city,
where the ship would not be spotted.

Adolph, the nineteen-year-old son of Margaret and Frederick,
met the vessel on the Long Island Sound at Rye. He then led the
nine disoriented slaves on foot across Westchester, from the Sound
to the shore of the Hudson. It must have been an incongruous sight:
Adolph and the nine Angolans, humping it overland in the trackless
New World night, the coddled child of two Dutch supermerchants
with these clandestine offerings from the troubled heart of Africa.
The captives must have been scared witless, raw from the slave gal-
ley, sea legs gone to jelly, sores on their ankles from the iron man-
acles, some of the taller, stronger men still bearing their shackles.
They were meeting for the first time the beruffled boy who could
easily be their master for the rest of their days.

They stopped their march at a barely cleared piece of riverbank at the Upper Mills. They were not anticipated there, although their keepers surely had begun to think about purchasing slaves to labor on their land. The presence of these particular captives was unpremeditated, resulting only from their unsalable condition and the fact that they somehow survived the months-long journey from Africa. (Later, customs officials would unsuccessfully prosecute Frederick for violating the English Navigation Acts with this backdoor smuggling fandango.)

A mill most likely stood on the Pocantico River when the prisoners arrived, fitted with the grinding stones Margaret had imported from Holland, awaiting workers. But there would be no sleeping quarters yet, no roof over their heads. The slaves would be expected to raise their shelter themselves, although their owners considered that a lesser priority than for them to begin clearing the fields of trees and rocks in order to plant wheat.

These men and women represented the nucleus, the elders, of the slave community that worked the vast Philipse lands in the Hudson Valley. (There was an exception: One Angolan, a man identified in a sailor's deposition as having a solitary eye, found himself dragooned into Frederick's service in New York City instead of at the Pocantico.) Frederick chose English names for his slaves— Symon, Charles, Hector, Susan, Mary, and Peter among them. The first generation probably bore the marks of their ethnic heritage, the most obvious being the shape of their teeth, which often were filed into diamonds, as was the tribal custom. None communicated in the tongue of this new land. When they learned the language of their captors, they would pass on their knowledge to the next wave of men and women to join them, brought by Philipse ships.

By the turn of the century, the family would hold a retinue of at least twenty individuals who served at both the Upper Mills and the Lower Mills in Yonkers, at a time when most New York colonists owned no more than one or two slaves. The horror of West Indian race relations did not touch Margaret and Frederick personally, though they put many Africans in the hands of island planters. But neither Margaret nor virtually any of her American contemporaries saw trafficking in humans as wrong. And that is why at Spring Head, a Dutch merchant's house that was almost certainly less ostentatious, more tasteful than some planters' homes, family members could gaze from their crest of land out to the glimmering Caribbean and contemplate the good fortune that followed them to all corners of the earth.

$\mathcal{S}$ometime in the early 1690s, Margaret's oldest son Philip met an Englishman's daughter named Maria Sparkes during one of his sojourns on Barbados. Maria had spent her formative years amid the island's sugar plantations, and family lore gives the Sparkes clan a burnished spot in the colonial history of the place, claiming her father to be one of its mid-seventeenth-century governors (though reliable evidence has never surfaced). But whether or not Maria's father held the questionably lofty job, the real power there surely belonged not to politicians but to the planters.

Philip's choice of a partner seemed ideal. In Maria he had found a young woman inured to the tropical airs that sickened so many settlers. She knew the potential for violence that suffused the island as thoroughly as its sugared air. The ratio of male to female colonists on Barbados had always tilted heavily toward the men, and if Philip intended to reside there he needed a wife who could tolerate the local discomforts.

Margaret had in all likelihood helped arrange this pairing, though her first choice probably would have been a Manhattan heiress. It was traditional for women of her station to orchestrate beneficial couplings for their children, and Margaret would hardly have left such important matters to chance. Annetje had her share of admirers, but had not yet made a match. Adolph had always showed more interest in commerce than in courting girls. Rombout has no presence in the historical record after the date of his birth—no references on ship lading, no land deeds or house mortgages, no political appointments, no court appearances, no mention in his father's will—suggesting that Margaret's youngest did not survive his childhood.

Eva, Margaret's daughter from her first marriage, had already claimed one of the esteemed Van Cortlandt brood for a fiancé. Just past thirty, Eva was getting on in years for a first-time bride. Perhaps she had been holding herself in reserve for thirty-two-year-old Jacobus, an up-and-coming merchant. He was the youngest of Oloff and Gertrude van Cortlandt's children. The sprawling Van Cortlandt compound was just down from the Philipse home on Stone Street. The five Philipse and seven Van Cortlandt offspring had grown up together, shuttling between the salt breezes and muck of the harbor and the grimy hiding places around Oloff van Cortlandt's old brewery. Because of those years of scrappy intimacy, and because they swam in the common liquid of uncommon affluence, the match seemed so easy.

Eva and Jacobus announced their intention to wed in 1690. The nuptials would take place in May 1691. Margaret arranged a choice wedding gift, an early dowry of sorts: In November 1690 she and Frederick settled upon the engaged couple a prime piece of New York City real estate: "All that his house and grounds situation lying and being on the East side of the Dock or mould of this Citty ffronting to the house that Mr. William Morris now lives in together with all and singular the Appurtenances thereto belonging or Appurtaining together with competent vallue of money or goods as the said ffrederick Phillips shall think suitable for the better preservation and maintenance of them."

In plain language, a harbor-front home, with all the trimmings.

The paperwork elucidating the terms of the gift omitted the signatures of both Margaret Hardenbroeck and Eva Philipse, benefactor and beneficiary, who now found themselves unalterably covert, invisible under English law. Nonetheless, Margaret could appreciate the sentiment embedded in the document's unwieldy wording. "Ffrederick" promised the gift in consideration of the intended marriage of the fiancés, "and for the tender love and affection which he beareth to his said daughter Eva," just as soon as the marriage was solemnized.

Jacobus came from venerable stock, and each surviving child of the Van Cortlandt family was considered fantastic spoils in the marriage lottery. Stephanus, Sophia, and Cornelia had all made worthy matches among the worthiest sector of Manhattan society: Stephanus to Gertrude of the eminent Schuyler clan, Sophia to Andries Teller, Cornelia to Brandt Schuyler. Maria van Cortlandt, Jacobus's much older sister and Margaret's sometime Albany partner in trade, had long ago become the doyenne of the great Rensselaerwijk estate.

Jacobus's widowed sister Catherine was, at thirty-nine, the last marriageable Van Cortlandt. She was only nine when she watched her sixteen-year-old sister Maria leave home to wed a much older man. Later, Catherine heeded the plea of her ailing sister and brought Maria's nine-year-old daughter to live with her in Manhattan and receive a New York education with an imported Dutch schoolmaster. At twenty-three Catherine had wed a prominent trader, a union that brought no children to complicate a future marriage. Her inheritance upon her husband's death comfortably padded the fortune she had received at the death of her father in 1684.

Margaret had linked one of her children to the Van Cortlandt clan, and knew that her eldest son Philip likely would not consider a well-off Manhattan widow for a spouse, now that he had found

Maria Sparkes. But perhaps twenty-five-year-old Adolph would one day show an interest in Catherine.

*O*verseeing the betrothals of her children was the last business Margaret pursued. In 1691 the she-merchant died at the age of fifty-four. Margaret's relative youth and the fact that she passed without a will suggest a sudden demise. Moneyed individuals of Margaret's era almost always made out wills. It would seem reasonable that she of the "terrible parsimony" would want to control her estate as assiduously in death as she did in life, if she had any choice in the matter. Since she died intestate, her extensive assets, property, ships, and lands devolved automatically upon her husband. Margaret's children would not receive any of her worldly goods until a decade later, when Frederick bequeathed them as he saw fit. The only exception was twenty-eight-year-old Philip. He had already proved himself a capable heir, and his parents had decided that Philip would be the proper Philipse to take over the Barbados business and the lovely Spring Head estate.

Funeral practices had changed only slightly in the three decades since Margaret laid her first husband to rest. In 1691 Conradus Vanderbeck and Richard Chapman were Manhattan's designated *aanspreckers,* now termed "inviters to the buryiall of deceased persons" and appointed and licensed by the English mayor. Margaret's family and colleagues would attend her corpse in the *doed kamer* of her home on Stone Street. In the Dutch tradition, she would have long ago stashed the requisite materials to ensure her own "good burial," the imported leather gloves and monkey spoons—gifts that would remind the guests just what a high muckety-muck she was.

On the day of the funeral her husband would set out pitchers of the chilled wine that was generally viewed as more elegant than the locally brewed beer and had the added benefit of being much higher in alcohol content. A cadre of strong men would lift the black-pall-draped coffin and march slowly to the graveyard. All the female guests would stay behind at Margaret's house, getting a start on that cold wine and the white Gouda pipes of Virginia tobacco that Margaret, in keeping with Dutch tradition, would have instructed her husband to offer her mourners. And after her body had been interred, when the party converged upon Stone Street and the ribaldry began, the men and women would dig in to the buttery, caraway-stuffed *doed-koecks*—as much a funeral staple among the English as they were among the Dutch. The she-merchant had specified in no uncertain terms that the cakes must be scored with the letters *M* and *H,* initials that would pass with her into her grave.

Part Two

1692–1783

CATHERINE

—

JOANNA

—

MARY

10

The Church of Catherine

*T*hrough the tropical forests of seventeenth-century Madagascar clomped the elephant bird, a portly, flightless giantess whose stature and solidity belied the fact that her species was in the last throes of its existence on earth. Three centuries later, only her fossil eggs would remain, the pale gray of cement and the size of volleyballs, now reduced to broken fragments carpeting the swollen dunes of the country's southern coast. But when European pirates descended upon the island to escape the law and count their gold, the largest bird that ever lived walked among them.

The beady-eyed creature was actually fairly shy. Ten feet tall, with a half-ton form seated heavily on leathery three-clawed feet, the bird preferred to stay put in the jungle shadows if at all possible. Still, no visitor who managed to catch a glimpse of the creature could forget it, even among the other marvels of Madagascar.

The island's flora and fauna resembled nothing a mariner from Europe or even America had encountered. The pristine rivers that flowed into the Indian Ocean from the island's gut carried minerals that stained the talc-white coastline the electric red of blood.

And through the open windows of the port cities, along with the slap of the waves and the clanking of ship metal in the harbor, newcomers could hear the keening cry of the indri, a lemur with lemon-drop eyes, and easily mistake it for the wail of a human toddler as it carried a mile or more across the hilltops of the rainforest. The Malagasy natives who populated the forest reported seeing a giant lemur that resembled a sloth they called a "tratratratra." They also told of a pygmy hippo and a panther-hunting cat-dog-rat that bore its young in the crown of the baobab tree. On the east coast of Madagascar, perpetual showers bathed an eight-hundred-mile escarpment. Baleen whales, humpbacks, and blues swam below. The trade winds swept through the bamboo groves and tree ferns. Farther inland the arthritic limbs of the ylang-ylang trees carried armloads of yellow waxen flowers.

For the pirates of the 1690s, Madagascar was nothing short of paradise. Bordering the southeast coast of Africa and twice the size of Great Britain, the island was like a continent, complete unto itself. There were sheltered coves all around the coast to hide ships, and beaches for careening an ocean-damaged hull—a laborious process that entailed scraping off barnacles, burning off seaweed, replacing rotten wood, and spreading on a protective paste of oil, tallow, and brimstone. Sailors found groves of the lemons and oranges they needed to treat scurvy. The geography seemed tailor-made for maximum piratical conquests, with equal proximity to the two major trading routes of the Red Sea and the Indian Ocean. The island offered a smorgasbord of provisions: oxen, goats, sheep, poultry, fish, rice, cotton, and sweet tropical fruits. Uncorseted native women offered themselves as wives.

Despite its charms, remote Madagascar probably would not have emerged as the foremost pirate haunt of the 1690s if it were not for a charismatic buccaneer named Adam Baldridge. Baldridge apparently hailed from New York and killed a man on the island of Jamaica before making his way to Madagascar in 1691. There he discovered the island of Sainte-Marie, which floated ten miles off the northeast coast of Madagascar like a pulsing satellite of opulent villainy.

Baldridge transformed Sainte-Marie into a pirate paradise inhabited by as many as fifteen hundred sailors at a time. A trading post he opened served up high-quality food and drink, and provided sewing needles, looking glasses, and other basics that even macho corsairs needed. In return, Baldridge demanded a share of the loot his customers acquired in the course of doing business. He declared himself "the King of the Pirates" and built a castle on a hill behind a stockade defended by forty guns. The pirate king acted as a go-between for competing Malagasy tribes and comforted himself with a harem of Malagasy girls. Baldridge also was the bridge between Sainte-Marie and England's American colonies. If any merchant wanted to do business with the pirates, Baldridge was the man to know.

Catherine van Cortlandt would become well acquainted with the bounty of Madagascar during the decade of the 1690s, more intimate than she ever could have imagined. Not the island's gigantic elephant bird eggs, yellow-eyed lemurs, or fragrant jungle bouquets. She would come to know Madagascar through an even more haunting product of the land—the Malagasy men and women shipped to Barbados and New York as slaves, a specialty commodity of her new husband, Margaret's widower, Frederick Philipse.

*C*atherine had been married to merchant John Dervall for thirteen years when he died in 1688. Frederick Philipse knew Dervall well—and was cognizant of the robust health of his estate. They had been partners in various business ventures over the years. Frederick had even suffered prosecution alongside Dervall on a smuggling charge in the 1680s, when both men were members of the governor's council. A scant year after the death of Margaret, his wife of nearly thirty years, Frederick must have felt a certain urgency to find a new mate to help manage his life—a life that included extensive commercial commitments, political appointments, real estate holdings, and four grown children.

So Frederick made his choice. On November 30, 1692, the sixty-six-year-old Frederick took forty-year-old Catherine to be his bride. To delineate the financial understanding that gave the union a bracing clarity, husband and wife put their signatures on an unsentimental, precisely worded prenuptial agreement of the sort that the Dutch community of New York still signed, despite English law and convention. Catherine kept the money she brought to the match. She alone controlled her personal fortune. Likewise, Catherine agreed not to make any legal claim on the Philipse empire. The agreement seemed more than adequate for this middle-aged bride, whose inheritance from her wealthy father, Oloff Stephanus van Cortlandt, had fortified the fortune from her first husband.

Catherine's economic autonomy probably figured in her new husband's satisfaction with their pairing, but her impeccable lineage naturally added quite a bit to her luster. Not only had Frederick done business with Catherine's first husband, he also had been a slightly younger peer of her father. Oloff van Cortlandt had arrived at Manhattan in 1638 and married Annetje Loockermans, an indomitable woman who, like Margaret, brought ceaseless energy to the settlement of New Amsterdam. Oloff, like his neighbor Frederick, was a political animal and had grown into one of the colony's wealthiest denizens. For his funeral in 1684, Oloff ordered a public display of his fortune: Each of the scores of mourners received a sterling spoon engraved with his full name. He also furnished each of his offspring with a private bequest that would set them up for the success a Van Cortlandt heir deserved.

Frederick also went back many years with Catherine's eldest brother Stephanus, one of the great achievers of his generation. Stephanus was appointed New York's first native-born mayor in 1677 at age thirty-six. Since 1675, Frederick and Stephanus had served together on the provincial council, the colony's reigning legislative body. But the strongest personal link was forged in May 1691, when Catherine's younger brother Jacobus—her favorite brother—married Frederick's stepdaughter Eva. The social events surrounding their marriage had thrown Catherine together with her future husband as never before. Margaret had died that same year.

Nuptial festivities in 1690s New York were usually extensive and drew on both Dutch and English traditions. Most weddings took place within the intimate surroundings of the bride's parents' home and could be performed by either a minister or a justice of the peace. An engaged couple no longer had to publish their wedding banns three times at a church. Instead, the groom must obtain permission to wed from the governor in the form of a marriage license (at a cost of half a guinea). At the wedding, a bride who cherished custom might wear a stiff, embroidered heirloom crown; the yardage of Asian silk incorporated in her cloud of petticoats proclaimed the amplitude of her dowry. The post-ceremony party typically lasted for days. The groom's parents usually sponsored the next-day open house, then the bride visited with friends and family for miles around, a round robin achieved on the back of a sturdy horse. Sack possett, the wedding beverage of choice, was a potent brew: Spanish sack, a dry white wine, stirred over a hot fire with sugar, eggs, and milk until the mixture became thick enough to eat with a spoon. One New York observer wrote a letter describing the average nuptial frenzy: "The Wedding-house resembles a beehive. Company perpetually flying in and out."

After such merriment the newlyweds deserved a rest, which they took in their own home. Some couples might have a rude surprise when they woke to the sight of a "May tree" outside their door, hung with mangled stockings, which would remain until the groom paid the pranksters to cart it away. More likely in the case of a second marriage such as Frederick and Catherine's, a poor box would be set out to collect good-works contributions, which they would donate to the church.

The marriage of mature partners often was born of practicality as much as romance. The pairing of these two almost could have been ordained by Margaret herself out of the desire to see Frederick and her children cared for and her memory undermined as little as possible.

It was only to be expected that Catherine would someday remarry at her social level. But still the timing was delicious for someone with a taste for advancement. Just one year after their wedding, in 1693, Governor Benjamin Fletcher granted Frederick Philipse a royal charter to create the "Manor of Philipseburg" out of the fifty-seven-thousand-acre tract he had assembled in the Hudson Valley, between Spuyten Duyvil and the Croton River, and between the Hudson and the Bronx River. King William and Queen Mary officially confirmed this elevation in status. The charter granted "our said loving subject" (in the overearnest governmental cadence of the day) a number of privileges, including the right to hunt, hawk, and mine. Catherine would be lady to her husband's lord, though both titles were strictly honorary and did not confer actual status in the English court.

The Philipsburg charter included permission to "erect a toll-bridge" at Spuyten Duyvil "and to charge tolls for its use." King William and Queen Mary had expressed interest in the project for years, but had not wanted to invest in a bridge. Now Frederick gained the right to assume the project as long as he charged "easy and reasonable" rates. In return for the privilege of running the enterprise, Frederick would pay the Crown a nominal four pounds twelve shillings annually.

Frederick chose the logical spot to build his bridge, at the shallowest section of the stream, the exact site of the age-old fording spot known to all as the Wading Place. The charter set forth the schedule of tolls (the king's troops and wagons and coaches exempted, of course). Since 1669 a ferry service had been in place on the creek, east of the Wading Place, operated by the enterprising Johannes Verveelen, who also opened a tavern on a meadow at the river's edge where he offered "three or four good beds for the entertainment of strangers." But travelers had avoided Verveelen's ferry tolls. They preferred to cross for free at the Wading Place, even if they occasionally got their feet wet. They grumbled even more loudly about the new bridge, about having to pay three pence just to get to Manhattan with dry boots—or worse, twelve pence to cross a score of sheep or hogs. The bridge toll could not be dodged with a private boat: Frederick charged nine pence each time the bridge had to be drawn up. Colonists complained even more about the actual procedure for passing over the span, which entailed banging on the gate and waiting—often a foot-numbing time in the cold and wet of winter—for the gatekeeper to open up and let travelers through.

With the construction of King's Bridge—the name the Crown imposed, according to "our royal will and pleasure"—the family now controlled the passage of every person or vehicle coming to the island of Manhattan from the north or going the other way. Frederick evicted Verveelen and took over management of his tavern. For anyone outside the family or the Crown, the King's Bridge represented an abrupt slide from freedom to constriction, with someone, no less a manor lord, keeping tabs on just who passed by, and when, and in whose company, and in what condition—a devolution that never has sat well with Americans, even in colonial times.

The heavy gate of the King's Bridge swung open freely for Lady Catherine. Like Margaret, Catherine often made her way up to the family's property from Manhattan on horseback, but she no longer had to worry about splattering her imported petticoats when she reached Spuyten Duyvil. The solidity and height of the bridge protected her. So did the clothing she wore. This lady's costume was all about control, about containment. Wrapped head to foot in her crimson capuchin, a hooded velvet cloak, Catherine would not go out to ride without also protecting her complexion against the elements. Like other women (and some men) of her class, she wore a "sun-expelling mask" designed specifically for that purpose, which resembled the harlequin mask of costume balls. Its stiffened silk—white or black or a more festive green or pink—had a silver mouthpiece to hold it in position, with two silk strings dangling silver beads that the wearer gripped in either corner of her mouth. With some practice, a lady could work the reins and even manage a somewhat lockjawed conversation without any mask slippage. This technically demanding fashion, imported of course from the Continent, was an accessory of widespread popularity at the end of the seventeenth century. For Catherine, passing over her family's bridge and trotting through the enormous acreage where she was suddenly more important and more visible than ever in her life, the mask not only served the cause of beauty, it also gave her a modicum of privacy.

When she married Frederick, Catherine found herself sharing dominion over a sweeping expanse of land, not just a manor, but an out-and-out empire. Even for an upright, sensible widow, the sensation had to be intoxicating.

$\mathcal{A}$t the time of Catherine's marriage, the atmosphere of New York and its environs embodied stark contrasts. Those with taste as well as coin indulged in wardrobes made from sumptuous textiles imported from Europe and the Orient—especially if they happened to be ship merchants like the Philipses. Fashionable clothes took a central place among the concerns of New Yorkers.

One of the earliest views of Manhattan shows two colonists in typical seventeenth-century garb, with beaver hats, seemingly engaged in the trade of goods including furs and tobacco. (A beaver lies on the ground between them.) It is the most accurate picture we have of what Margaret Hardenbroeck and Frederick Philipse might have actually looked like. The horizon offers some important New Amsterdam landmarks: The military fort, signal flag, gallows, and small houses can be seen clustered at the tip of the island.

An 1894 engraving of New Amsterdam's Heerengracht Canal in the middle of the seventeenth century. While the artist might have slightly romanticized the scene—the canal was almost surely never as tidy and clean as it appears here—he gives a good sense of the atmosphere of the spot as a place to meet and socialize.

The first street plan of Manhattan was created for the Dutch West India Company in 1660. This 1916 copy was enhanced by the artist in order to clearly depict the details of life on Manhattan, including a representation of every street, dwelling, warehouse, and garden as well as ships anchored in its harbor and even some of the boulders in the shallows offshore.

Collection of The New-York Historical Society, neg. no. 57812

This painting depicts a typical Holland interior, in which a mother with newborn twins receives assistance from two attendants, possibly family members. The first women to settle in New Netherland brought the domestic customs of the Old World with them, including the traditions of birth and child-rearing.

Courtesy of Sotheby's Picture Library, London

A robust example of a utilitarian *kas,* the oversized standing cupboard that held pride of place in New Netherland households.

Photography Collection, Miriam and Ira D. Wallach Division of Art, Prints and Photographs, The New York Public Library, Astor, Lenox and Tilden Foundations

The streets of Amsterdam were filled with she-merchants, such as this late-seventeenth-century Dutch fishwife, when Margaret Hardenbroeck was growing up there.
Rijksmuseum Amsterdam

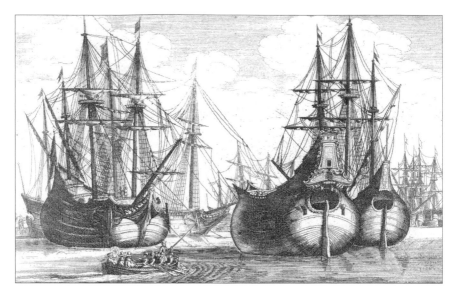

Margaret's fleet included several of the little ships called *fluyts*, the workhorses among Dutch merchant ships. In this seventeenth-century depiction, note the swollen shape of the vessels, whose revolutionary design allowed merchants to pack in as many goods as possible.
Library of Congress, Prints & Photographs Division, LC-USZ62-74941

Notarized acknowledgment of debt by Margrieta (Margaret) Hardenbroeck, January 9, 1664. This was a standard legal record in the trading transactions of the time.
Amsterdam City Archives

This lithograph was based on a sketch in Jasper Danckaerts's 1679 journal describing his trip to America. Depicted from a vantage point north of New Amsterdam, the scene looks as it might have to Margaret when she traveled south from Yonkers to Manhattan.
Emmet Collection, Miriam and Ira D. Wallach Division of Art, Prints and Photographs, The New York Public Library, Astor, Lenox and Tilden Foundations

The King's Bridge (seen in the distance in this 1860 illustration) was built by Frederick Philipse I in 1693 and gave free passage to Catherine Philipse and the rest of the family over treacherous Spuyten Duyvil creek, while others had to pay three pence to cross. In 1759 the Farmer's Free Bridge was built to give free passage to all travelers. The King's Bridge stood until 1913, when Spuyten Duyvil was partly filled in.
Picture Collection, The Branch Libraries, The New York Public Library, Astor, Lenox and Tilden Foundations

A "sun-expelling mask" imported from the Continent offered high-fashion
protection against the elements for women like Catherine Philipse.

This 1850 watercolor shows the Philipsburg Upper Mills on the Pocantico River as it might have looked in Catherine's day.
Historic Hudson Valley, Tarrytown, New York

The Dutch Reformed church at Philipsburg is little changed since Catherine oversaw its construction in the 1690s, aside from the reorientation of its entrance from the south wall to the west at some point before this 1864 photograph. The bronze bell Margaret imported from a Holland foundry in 1685 still calls parishioners to worship on summer Sundays.
Rare Books Division, The New York Public Library, Astor, Lenox and Tilden Foundations

This map was created in 1735 by a Manhattan resident who signed her name, Mrs. Buchnerd, in the upper left corner. Not a professional cartographer, she highlighted those personal landmarks she favored, including "coffy houses," pleasure gardens, and more than one "fishing place" along the island's shore.

Joanna Brockholes by John Wollaston, ca. 1750, oil on canvas, 20 x 25"
Colby College Museum of Art, Gift of Mr. and Mrs. Ellerton M. Jette, 1982.015

Frederick Philipse by John Wollaston, ca. 1750, oil on canvas, 30 x 25"
Colby College Museum of Art, Gift of Mr. and Mrs. Ellerton M. Jette, 1982.014

New York's slave revolt of 1741 never brought down the white establishment, but the official investigation into its origins resulted in mass executions: Twenty-one men and women were hanged, thirteen were burned at the stake, and seventy-two were expelled from the colony. One of the first to die was the Philipse slave Cuffee.
Picture Collection, The Branch Libraries, The New York Public Library, Astor, Lenox and Tilden Foundations

Mary Philipse was a twenty-year-old ingenue when John Wollaston painted her likeness in 1750.
Historic Hudson Valley, Tarrytown, New York

As a matron of forty-one, Mary Philipse Morris commissioned this 1771 portrait by John Singleton Copley, the preeminent society portraitist of the era.
Courtesy, Winterthur Museum

In September 1776, after the British marched in to occupy Manhattan, a fire presumed to be the work of Patriot arsonists burned through a quarter of the city, leaving many residents homeless and decimating some of the city's landmarks, including Trinity Church. The flames also consumed the home of Roger and Mary Philipse Morris.
Library of Congress, Prints & Photographs Division, LC-USZ62-50093

Philipse Manor Hall as it appeared in the middle of the late nineteenth century, past its luxurious prime—note the stone wall and city sidewalk—but still an impressive residence.
Emmet Collection, Miriam and Ira D. Wallach Division of Art, Prints and Photographs, The New York Public Library, Astor, Lenox and Tilden Foundations

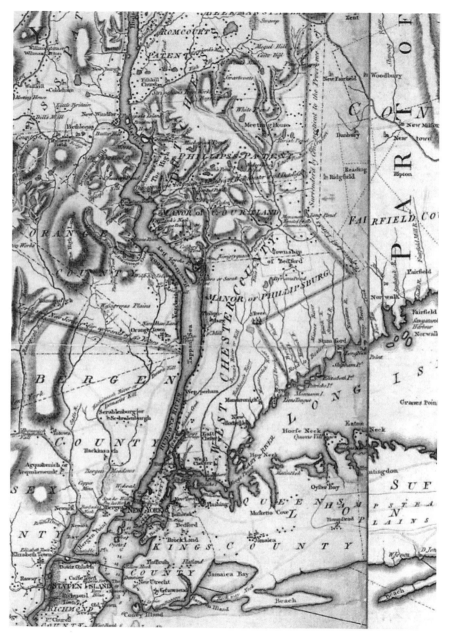

A Revolutionary War–era military map indicates some of the landmarks of Philipsburg Manor.

The Lionel Pincus and Princess Firyal Map Division, The New York Public Library, Astor, Lenox and Tilden Foundations

At the close of the Revolution, an unknown artist depicted Philipse Manor Hall and its environs, probably for use as a real estate circular in the wake of the property's confiscation, division, and sale by the new American government. It is the one surviving picture of the manor from before the turn of the eighteenth century.
Historic Hudson Valley, Tarrytown, New York

A lady of Catherine's time, parading down Wall Street or Broadway, would wear a dark high-peaked felt hat and sheer scarf made of the soft, pure semilustrous silk called lutestring. Her glazed petticoat, quilted in fanciful designs, might offer a splash of scarlet, but the silk gown worn over it typically was styled in the conservative, muted colors considered more acceptable for outerwear. A pure gold chatelaine—essentially a heavy chain with some utilitarian value—hung from every respectable woman's waist. From it she suspended the necessities of her day, including a thimble, scissors, and keys to the many strongboxes and locked cupboards in her home; hoarding the family's valuables fell as often as not to her. She stored gourmet tea leaves in boxes of mahogany and gilt brass. Even spices and wax candles were hoarded. Both items were scarce and expensive, in a time when both the flavor of food and the quality of light had an importance impossible to exaggerate.

Wealthier families had accumulated stylish goods for the home that showed their prominence and sophistication. Many of these accessories and furnishings originated in England, and others represented exports from the extremely trendy Far East, making them quite different than the finest Dutch items of earlier decades. A well-to-do household might prominently display a heavy, handblown British hourglass. Wristwatches were still nonexistent—even as late as the American Revolution, George Washington carried a pocket sundial as his timepiece. A merchant's table might hold a pair of nippers for the cone of sugar the family was lucky enough to afford, along with a chocolate pot—to be distinguished from a teapot, coffeepot, or hot milk pot by its shape and the position of its handle—and china cream jugs and spoons of carved horn.

A hostess of Catherine's stature would be proud to produce something extravagantly rare: a knife individually assigned to each dinner guest. The household's polished satinwood knife box, with its velvet, padded slots to hold the ivory-handled knives or those with handles of white split bone, stood on the sideboard as a status object in its own right. And yet forks had not yet entered common usage, even among the rich. After they cut food with their expensive knives, the most refined New Yorkers shoveled meals to their mouths on a dripping wedge fashioned from their reliably unwashed fingers and thumb.

It served as the perfect metaphor for the omnipresent flip side of the house-proud, clothes-conscious Dutch-English colonial lifestyle of New York. A society whose very existence was premised upon the killing of huge numbers of forest animals (and thousands upon thousands of skins still reached Manhattan from the upper Hudson in a single year) necessarily encompassed, even celebrated, the grittier dimensions of life. For Margaret, it had meant shaking the hands of the people who did the killing, and touching with her own fingers the brain-tanned hides of forest creatures to check the skins' quality. Even for Catherine Philipse, at the pinnacle of New York's social hierarchy, careful opulence coexisted with Brueghelian earthiness.

A very basic example: Catherine could afford the most marvelously intricate gowns imported from Paris. No woman, though, not even Catherine, would dream of shielding her nether regions by pulling on a pair of underpants, even when she menstruated. Women simply bled into their clothing—we're talking about roughly thirty years of monthly "accidents," except for the months a woman spent pregnant. All female colonists shared this condition, of course, not just New Yorkers, but Catherine's blood soiled a wardrobe that featured astronomically priced, handcrafted gowns. Perhaps the practice of ignoring the issue had its advantages: One historian surmised that far from finding menstrual blood a turnoff, men of the era perceived the aroma of a woman's monthly flow as intensely seductive. And that is fortunate, since bathing with soap and water still was actively frowned upon, with the inevitable gaminess ameliorated mainly by sachets sewn into clothes linings.

Lady Catherine inherited Margaret's rubblestone house, surrounded by Hudson Valley wilderness and still consisting of two
unadorned rooms and a basement constructed for the down-to-
earth business of shipping skins. Buckets of drinking water had to
be hauled laboriously from the Nepperhan River, and light at night
came from the canopy of stars above. Despite the predilection of
the Hudson Valley Dutch for stuffing beds into rooms, and family
members (and friends of family) into beds, a two-room house
meant cramped quarters for six to eight well-padded aristocrats.

That basement also spent some time as a backwoods bank, according to legend. After all, it was easy enough for smugglers'
sloops anchored down by the Strand to deliver contraband or the
silver Spanish currency pieces of eight. Some said a tunnel from the
shore cut indirectly to the dim, green-tiled room beneath Margaret's *groot kamer*. The cellar that now belonged to upstanding
Catherine would have held stacks of coins, each bearing the image
of the twin pillars of Hercules, at least some of them procured
through illegal activities on the high seas.

Manhattan was gripped by pirate fever at the turn of the seventeenth century. In the early 1690s, New York's Governor Fletcher
not only turned a blind eye to the pirates' presence in Manhattan
but actively encouraged their settlement on the island by selling
amnesties for criminal acts for next to nothing. Practically every
little cove on the East River sheltered a ship bearing the Jolly
Roger (or, just as often, a personalized flag adorned with crossed
swords, dancing skeletons, a bleeding heart, or an hourglass, with or
without a skull). Pirates openly frequented the taverns of Dock
Street, spending their pay on rich meals. Bulk spices and bolts of exotic fabrics from the East materialized in certain merchants' warehouses after pirate ships came in to berth in the East River harbor,
and few questions were asked about their provenance. After dark,
pirate gangs roamed the maze of wharves and docks to appropriate

the rope, sails, brass, iron fittings, and even anchors from ships and storehouses. Yet Manhattan welcomed the sailors, who spent lavishly at any establishment they patronized and pumped into the local economy nearly the equivalent of what they took out.

There was a fine line between legitimate and illegitimate activities on the high seas, especially during the early heyday of pirating. After all, every powerful monarchy hired thoroughly reputable captains to cruise the ocean as privateers, happy to ignore the seamen's tactics as long as they split the spoils with the Crown. Privateers frequently jumped the line to pirate. It was a transition made famous at the turn of the century by the case of William Kidd, a Scots Presbyterian minister's son who made Manhattan his home port and appeared to flame out after being hired by the English admiralty to chase "Pirates, Freebooters, and Sea Rovers" in the Indian Ocean. When the authorities hung his body to rot in chains at London's Tidbury Point as an example to would-be pirates, he left behind a wealthy widow named Sarah Oort, a respectable town house on Pearl Street overlooking New York Harbor, and a community astounded at how such a regular guy could have gone over to the dark side in such flamboyant fashion.

The pirates of Madagascar hatched a society that ran counter to the monarchical culture of England in dramatic ways. All hands on a pirate vessel consumed equal rations, unlike the hardtack-crunching swabs on a Philipse ship who watched the skipper and his stateroom guests put away fresh-killed chicken and aged Madeira. Pirate crews actually elected their captain by majority vote, along with the quartermaster, and either could be deposed if the majority grew dissatisfied with their leadership. The crew, not the captain, determined destinations, ships to attack, or coastal villages to raid. Most important, they divvied up all plunder according to a standard they themselves devised (captain, quartermaster, master, boatswain, and gunner received larger shares, and crew members who lost a body part in battle could count on appropriate compensation).

Violence pervaded the business of raiding ships, no matter how democratic the outlaws' chain of command. Pirates cut the throats of skippers. They tortured victims with whacks of the cutlass, lit matches held to eyelids and fingertips, and by "woolding," which meant twisting a slender cord so tightly around the skull that the eyes popped out. A Spaniard who survived a 1668 attack by Henry Morgan singled out the treatment of female prisoners, who were "burned in parts that for decency he will not refer to." Flogging, flaying, and brining (and sometimes all three) were standard fare.

But the racy stench of piracy was not going to make one of the foremost traders of the day hold his nose. Where others might have seen brutality, Frederick Philipse saw opportunity. True, this lord of the manor never stormed a ship, never massacred a crew, and never holed up with a fortified beverage at a private pirate haven. But after twenty-five years as a respectable politician, a longtime member of the governor's council, and an upstanding member of the New York elite, the ship merchant's boundaries between legitimate and illicit international trade had blurred. To give him credit, Frederick only entered the business when he got what amounted to an engraved invitation—a 1691 letter from Adam Baldridge, Sainte-Marie's resident pirate king. Baldridge proposed a transaction that would furnish Frederick with two hundred premium Malagasy captives at thirty shillings a head, well below the going price for African Gold Coast slaves. For any merchandise Frederick might want to unload on Sainte-Marie Island, Baldridge could offer pieces of eight, India goods, or whatever currency suited the ship merchant best.

Baldridge's proposal conveniently answered what Frederick considered to be a pressing need. Since that first cargo of Angolan men and women reached Barbados and then Philipsburg in 1685, Frederick found it increasingly difficult to obtain slaves from the west coast of Africa. England's Royal African Company had managed to keep out nearly all independent traffickers. Frequent rebellions also resulted in loss of life and a loss of revenue to the slave trade. Baldridge promised to provide significantly cheaper product if Frederick sent his ships around Cape Horn, up to Madagascar.

For the first several years of Frederick's Madagascar venture— also the first several years of his marriage to Catherine—the enterprise looked promising. The risk seemed minimal. The returns from a single voyage were much, much higher than other merchant activities. In a letter to his supplier on Madagascar, Frederick wrote, "Negroes in these times will fetch thirty pound and upwards by the head, unless they bee children or superannuated. It is by negroes that I finde my cheivest Proffitt. All other trade I look upon as by the by."

True, a few kinks needed to be worked out. In 1693, responding to Baldridge's invitation, Frederick sent the *Charles* on a round-trip voyage to Madagascar. The pirate king fell down on his word, delivering only thirty-four slaves—fifteen of them toddlers—and one hundred pieces of eight, rather than the two hundred slaves and one hundred pieces of eight he promised. But Frederick did not lose faith in this new venue. He wrote his fellow merchant to politely chastise the man, then in 1695 dispatched the *Katharyn* and a bark called the *Margaret* (a new vessel, not the one that belonged to *the* Margaret) off to Sainte-Marie. Records show that Frederick soon collected nearly ten thousand pieces of eight per voyage. A year and a half later, the *Margaret* returned to the port of New York with a load of slaves assembled not only from Sainte-Marie but also from the Madagascar mainland and Delagoa, at the southern tip of Africa, where glass beads, brass neck collars, arm rings, and looking glasses went a long way with African slave lords.

Serious traders also knew to send plenty of rum, which was in high demand and cheap to brew, especially for a sugar importer with tight Barbados connections. The *Margaret* set sail in 1698 with sixteen casks of rum stowed alongside hogsheads of salt, Madeira, beer, lime juice, and gunpowder. Like the Indian fur traders before them, the African slave traders of the Gold Coast developed an almost instantaneous addiction to alcohol, weakening their ability to drive any kind of bargain. Soon, the worth of Africans sold to European traffickers could be measured literally in gallons of premier New England rum. The devotion to rum was not limited to native rulers and slave dealers, however. Ship captains, crews, factors, governors, merchants, and pirates guzzled it, too. (Adding to its draw, rum was increasingly perceived as easier on the liver than brandy.) Tobacco also exerted a pull on the African dealers, who tasted the golden herb when white sea captains arrived in the 1600s. By early in the next century, coastal Africans were happy to hand over captives under their control to acquire the poor-grade, molasses-soaked Brazilian tobacco they now craved.

For now, Frederick did not have to worry that his slaving activities would affect his high place on New York's social ladder. Nothing had changed on that score since Margaret welcomed her first cargo of Africans to Barbados fifteen years earlier. A colonial merchant in the seventeenth century could invest in slave trafficking with impunity. Quakers alone objected, and they themselves were so despised that their criticisms had little weight.

Catherine along with the rest of New York society knew that her husband was heavily invested in slave trade. His vessels made their way out of the East River roadstead on a regular basis with an itinerary that was no secret. The couple's closest peers hailed the ships that returned from Africa with consignments of slaves they had ordered. These well-to-do buyers had no more need for slave labor than less prosperous New Yorkers did, at least in the way that rice and tobacco planters of the South claimed to need slaves for large-scale farms. But if you were a New York Livingston or Schuyler, the argument might go that there were complex mechanics in keeping various town houses running smoothly, requiring additional labor, and that European immigrant house servants were famously hard to keep. Affluent New York at the end of the seventeenth century had new British-influenced domestic rituals, a focus on elegant manners that a well-trained staff of liveried Africans could help maintain. Soon the practice of keeping a good-size stable of slaves became an aristocratic convention.

Enslaved Africans were bought and sold on wharves along the New York waterfront, as well as out of taverns. Colonial magnates were precise in requesting specific characteristics in their chattel: Gerard Beekman sought to purchase "a young negro wench and child of 9 months," and Cadwallader Colden explained that he needed to find a thirteen-year-old female slave because his "wife desires her chiefly to keep the children and to sow . . . [preferably] one that appears to be good natured."

In the summer of 1698, after Frederick's fleet had been plying the waters between New York and Sainte-Marie for five years with no interference, the British East India Company obtained a charter from the Crown that gave it a monopoly on trade in the waters around Madagascar. The British advised pirates that their business was no longer welcome in the seas surrounding their cushy stronghold in the Indian Ocean. As for Manhattan, the pirate haven across the seas, the Crown no longer would tolerate illegal activities there—a complete reversal of the blind eye it had cast in the past. England ordered its newest governor in New York, Richard Coote, first Earl of Bellomonte, to crack down not only on pirates but also on those who provisioned them and bought their purloined goods.

It seemed like the logical time to get out of the business. But the man who had ignored the requirement of an escort through Indian territory thirty years earlier refused to abandon a choice trade route when there was still money to be made. Frederick brokered one last deal in 1698—to bring a load of seventy Malagasy slaves into New York Harbor—and got caught.

Frederick did not swing from the gallows for his association with pirates, but he was forced to resign from his high-prestige post on the governor's council. The family also lost the valuable trade goods and slaves seized by Britain's High Court of Admiralty, which furthermore impounded one of Frederick's finest trading ships, the *Margaret*. Catherine could do nothing but share her eminent husband's shame before all of New York society.

*T*he Van Cortlandt children had been indoctrinated with the concept that the Lord would reward not only their piety but also their hard work. Oloff, one of the richest men in the colony, had insisted that the children help out in the family's brewery. These formative lessons gave the boys a vigorous nudge into trade and politics, fields they dominated.

It was a bit different for the girl who came smack in the middle of her successful Van Cortlandt brothers. In 1665 thirteen-year-old Cat, as she signed her letters—"With hearty greetings"—had gone to stay with Maria after the birth of her older sister's first baby and the advent of the septic arthritis that would cripple Maria the rest of her life. In the innocent ripeness of her own adolescent health, young Cat watched her sister raise seven children and gut out the management of household matters on her million-acre estate even when suffering wincing pain. That was a lesson.

The new lady of the manor also had learned from Margaret. Catherine and her siblings, playing in the Road, could not help but notice as the she-merchant strode toward the harbor to whip a ship's crew into shape or weasel an extra measure of wampum out of a two-bit deal. Margaret had invited Catherine and her siblings into her home to visit with the Philipse children, but on those occasions Margaret was not a mother who would bring out offerings of apples or pretzels. She was never unfriendly, but she had more important things to do.

Margaret Hardenbroeck and New Netherland's other women pioneers, the maidens who disembarked from the first ships at Manhattan—Catherine's own mother included—had a certain tough, purely Dutch way about them. Their determination to succeed in this New World was based partly on wanting to see their families thrive, but it was just as much for the sake of triumphing in and of itself. They were American facsimiles of the women who were the backbone of Holland.

Catherine's peers, the next generation, were different. They came to maturity just as the stakes changed. Catherine celebrated her twelfth birthday the year the English took New Netherland. For decades, Manhattan was still a Dutch town and the elders were Dutch, but an English flavor gradually permeated her life. It was not only the superficial, gradual transformation of manners from the Dutch style to the English. There was a more fundamental shift, an erosion of the underlying principle that women could participate fully in public life. Catherine would never participate in the world of large-scale buying and selling, courtroom dramas and important contracts. The earlier generation's example was the only vestige of that egalitarian era, a lifeline to a reality that soon would exist only in legend. This was the education by example that Margaret, Dutch-born she-merchant, passed on to Catherine, first-generation American, product of English laws and culture, whose profession was to be, essentially, a wife.

Sometimes it takes time for lessons to sink in. Twice married, middle-aged, and with money to spare, Lady Catherine finally realized that, like her predecessor Margaret, she had something more important to do.

Catherine, like Margaret before her, too, would build a house. Only hers would be a house of worship to shelter and sustain a far-flung community of the devout. (And perhaps even serve as a sort of spiritual amnesty for whatever sins she vaguely perceived in her husband's activities.) Everywhere in America, dutiful colonists built churches, but these early structures often fell short in materials or design or both, and all too soon passed into disuse. When Catherine muscled her project ahead to completion, the resulting edifice was so sturdy it would endure for more than three centuries.

Catherine built her church to last.

The Protestant Dutch Reformed church at the Manor of Philipsburg would be sited fifteen miles north of the house Margaret built, at the land parcel called the Upper Mills. There, a gristmill stood at the junction of the Hudson and the swift-running Pocantico River. A waterwheel turned the stones to grind the grain, with the flour packed in barrels on site, then loaded onto ships that sailed down the Hudson to transfer their contents to oceanbound trade vessels. At the mouth of the Pocantico, a bay pushed in nearly half a mile from the Hudson like a liquid thumb to actually touch the millpond. At some spots the bay had a width of more than one hundred feet, allowing sizable ships to sail all the way inland from the Hudson and tie up right beside the wharf to receive cargo.

Frederick and Margaret had talked about erecting a church in the area, likely with the idea of satisfying the new settlers and attracting others. Only four families settled in the region of the Upper Mills throughout the 1680s. A similar number of farms surrounded the gristmill at Yonkers. Then, in the 1690s, another twenty families came to Philipsburg. Despite the incursions of English settlers in Manhattan, the British were the exception in Philipsburg, whose colonists came mainly from Dutch, French, and German stock. A Reformed church might prove a draw for further settlers, which the estate desperately needed if it was going to prosper.

Margaret had never followed through on the church plans. At one point, she commissioned an elegant church bell from a bronze foundry, one of her Holland vendors. The bell was inscribed with the date 1685, and Margaret had it shipped all the way to Manhattan and up the Hudson in its straw-stuffed crate, to be off-loaded and admired at the mill wharf. But for ten years it sat in its wooden box.

Since the bell landed, the population of Philipsburg had grown to upwards of 150 colonists. These were hardworking people, eager to cultivate the land as tenants and make something of themselves. But the spiritual climate seems to have been profoundly ungodly. Westchester was "the most rude and heathenish country I ever saw in my whole life," wrote Caleb Heathcote, lord of Scarsdale Manor, "Sundays being the only time set apart by them for all manner of vain sports and diversions." Traveling ministers, he asserted, pocketed the collection money after delivering sermons "to please . . . and delight the fancies of their hearers." The time was ripe to assemble a congregation and appoint a regular minister. Catherine was ready to complete Margaret's work.

The church site, atop a bluff that rose between the trail that would become the Albany Post Road and the Hudson River to the west, made sense for a number of reasons. Local people already buried their dead in a shady grove there. It was a simple place, with small slabs of wood for markers, and many graves with no markers at all. No stone carver had as yet moved into the neighborhood; cutting the simplest initials into a chunk of rock stymied the average farmer.

River breezes cooled the hilltop, even on the stickiest August days. The Pocantico flowed past on its way to power the mill that put food on everyone's table. The church site's proximity to the hub of the milling operation was another bonus. Worshippers knew the route to the weather-streaked wood of the gristmill; they often traveled there by boat or by oxcart. And the site lay a relatively short stretch from the Philipse family's home on the Nepperhan, reachable by horseback in only an hour or two. Catherine could also take a sloop upriver on the Hudson, just a bump from one tributary to the next, disembark and stroll across a half mile of woodland to get to services.

The lady of the manor lacked specific knowledge of stone craft, but she chose the most experienced craftsmen available to put down each course of stones. To ensure it was done right, she personally supervised the work, riding up from Philipse Manor Hall or even from Manhattan on horseback—often mounted, according to local legend, on a pillion behind brother Jacobus. The pillion, sewn of buttery deerskin, was a bolster that assured a smooth, ladylike ride. A woman who used one was elevated, regal. Yet she was inhibited, of course, to some degree by riding sidesaddle— typical for colonial women—and more so by not actually handling the reins or commanding the horse, something Catherine certainly did on some occasions. It was an irony—the foreman of one of the major construction projects of its time being lofted delicately sidesaddle to the construction site—but not one that was unusual for the day.

Building any colonial structure was a science, but it began as a quest. Catherine decided to build the church with stone, which was more economical than brick but would not burn down like the cherished wooden medieval churches reduced to ashes in the 1666 Great Fire of London. And using stone made the quest easy. In the Hudson Valley, it was not necessary to quarry or cut anything. Instead builders used stones disgorged from the crumbly depths of loam as farmers cleared their land. The fields yielded more stones than any builder could ever use, no matter how many buildings or walls were wanted. The challenge was getting them from one place to another—in this case, from the low-lying river valley up a winding hill. The bluff above the Pocantico was somewhat less picturesque if your task involved hauling a one-hundred-pound boulder up its side. Catherine would solve that problem by sending her slaves to do the job.

Next came the science. It was necessary to literally roll a stone along the ground, as if you were rolling dice in a game. The side that landed facedown was its flattest face, the bedding plane, which would sit cushioned in mortar. The second flattest face, then, was the one to expose on the building's exterior. Selecting the stones for the project required the seasoned, perceptive hands of a master stonemason, with proper positioning required not only for aesthetic reasons but to ensure structural integrity.

For each exterior wall, the crew had to build a parallel interior wall, and fill the space between the two walls with a thick batter of stones, rubble, and mortar. After every couple of courses they set a bondstone that stretched the width of the interior and exterior walls, linking them into a sturdy mass. Workers could build no higher than five feet in a single day or the wall might collapse. Stones had to be wetted before use, which meant vessels of water also hauled up the hill. The masons, joiners, and carpenters who put together Philipsburg Church had the skill to construct the walls so that they were plumb on the outside but beveled on the inside—fully four feet wide at the base, curving in to twenty-two inches at the roofline—to bear the enormous weight of all that stone.

From clay pits on the banks of the river came gallons of mud, also to be slogged uphill. Hudson River mud had a supple texture that made it a superior base for the mortar that would permanently bind the structure together. Colonial builders evolved their individual formulas for mortar (along with formulas for house paint, whose volatility meant it had to be blended on-site and used within hours), which might include oyster shells and even eggshells. One 1703 recipe for "cold cement" called for grated cheddar, a pint of cow's milk, and a dozen egg whites as well as half a pound of unslaked lime. Frost rendered even the finest freshly applied mortar useless, so building started only in the settled weather of late spring.

Inside the church, that miraculous river mud found another use, smeared over every interior surface and left to dry before the conventional coat of a whitewash. The magnitude of the West Indian shipping enterprise at Philipsburg meant crushed, lime-dense Caribbean coral was readily available for the whitewash deployed on every manor surface, including the radiant walls of the church. When candles provided the only illumination on long, dark winter afternoons, the brightness of whitewash took on added importance.

By 1697 Catherine's work was done. The church bell that Margaret had brought from Holland finally rang with its call to worship on the bluff that rose above the mills, farmhouses, woods, and fields of Philipsburg. Inscribed on its side was a verse from Romans, *Si Deus Pro Nobis, Quis Contra Nos?*: "If God be for us, who can be against us?" Now congregants and every traveler on the road from New York to Albany could lift their gaze to the roof of the church and see the copper pennant that floated high above it, an artifact upon which a pair of imprinted initials was clearly visible: VF, for Vrederick Flypsen.

This is what Philipsburg Church looked like at its completion: a neat stone exterior with a gambrel roof. Seven clear square casement windows set high in the walls, crisscrossed by iron bars, with sills a yard deep, trimmed in quoins of yellow brick. Colored glass existed in New York—the Dutch Reformed Church of Albany had tinted windows inscribed with the names of bigwig donors—but not here in the still primitive Hudson Valley. The windows of the church Catherine built, though, were placed directly across from each other so that an aspect of sky met parishioners' gaze on all sides. The building had a wood-paneled ceiling with rough-hewn beams. Swallows twittered overhead. The congregation entered by a path at the south side and sat facing away from the broad dirt highway that some of the settlers had traveled to get there.

Hierarchy governed seating in the church. On a Sunday morning in the year 1700, say, Catherine would file in first, her countenance sober beneath her linen cap, her Bible suspended from her neck with a gold chain, a young slave behind her carrying a foot stove. The stove's iron and wood enclosed a metal canister containing coals from the hearth at home, and its hot steam rose from beneath her skirts for the length of the service, providing the only semblance of warmth in the building. She proceeded to her ornate curtain-cloaked throne, elevated above the rest of the congregation, which along with Frederick's throne flanked the pulpit. Entering the church after Frederick and Catherine, the yeoman farmers and craftsmen took their places with their wives on backless benches. The single men and boys sat in the upstairs gallery to the side of the southwest door, followed by the contingent of Philipse slaves. A choir filled another balcony opposite the door, under the watchful eye of the precentor—the a cappella rendition of psalms was the only music heard; no organ was customary in the simple Reformed sanctuary.

Guilliam Bertholf, the new dominie, traveled all the way from Hackensack every week to deliver the sermon. He climbed a winding staircase and stood beneath a hexagonal sounding board of carved mahogany that hung down from above and amplified his speaking voice. An hourglass sat squarely before the clergyman. The rule was that he could upend it just once during the service, no matter how impassioned he might become, which limited the session to under two hours.

Catherine could now sit back and enjoy the fruits of her labor, an upstanding lady married to a distinguished lord. During the time her church was under construction, Frederick had gotten more deeply involved in the Madagascar trade, sending provisions across the world to his all-consuming pirates and bringing back Malagasy men and women for the equally voracious New York slave market.

When the church opened its doors, it was 1697—the year before the great merchant's career would be cut short by his disgrace.

If God be for us, who can be against us?

*W*hile Catherine completed her church, her eldest stepson Philip finally married the privileged Maria Sparkes on the island of Barbados. Now thirty-three, Philip would bring his bride to live with him at Spring Head. The tropics were nothing like New York, Philip's hometown, and he and Maria were far from the support of his father and siblings. But managing the estate would allow them to create their own fortune separate from the monies each received from wealthy parents. They could develop a second-generation merchant enterprise, and start a family.

Success would require a constitutional hardiness as well as hard labor on this beautiful island plagued by illness, hurricanes, and periodic slave uprisings. For this couple, indolence had little attraction, as tempting as it might have seemed considering their material comfort. Philip had been working (and adventuring) since he was a teenager. Maria came to Barbados when she could have remained more comfortably at the family estate in Worcester, England, as did two of her sisters. Her choice to relocate to this often harsh and frequently ungenteel culture showed an independence that must have appealed to a man raised by an indefatigable mother.

When Maria bore a son the following year, the couple christened the infant Frederick after his paternal grandfather. Philip Philipse was the first male of his family's generation to marry, and the first to produce a child. (Eva and Jacobus van Cortlandt had one son already, but she was an adopted Philipse.) The baby's ambitious parents had every intention of raising and educating young Frederick II as the first of a healthy brood of second-generation Philipses.

It wasn't to be. Waves of disease had always pounded colonial-era Barbados. In one year, 1694, a single parish on the island buried 354 of its residents, while baptizing only 40. Yellow fever stalked the people of Barbados, striking without regard to victims' color or gender until the stink of death nearly smothered the scent of the cane vats.

Wealthy landowners did not escape. One witness to the contagion described his privileged neighbors: "Many who had begun and almost finished greate sugar workes, who dandled themselves in their hopes, . . . were suddenly laid in the dust, and their estates left unto strangers." Often, a planter barely out of his teens, having sailed from England to make his fortune, amassed an estate of tens of thousands of acres and dozens of slaves before succumbing to the island's killing fevers at the age of thirty or so. Funerals became fall-down-drunk spectacles where mourners remembered the dead with lavish feasts and took home souvenir dress hats and rings of gold. Stonecutters received top dollar to incise tombstones with the privileged victims' prized heraldic coats of arms.

In 1700, when Frederick II was just two years old, his father, Philip, became one of the fever's more prominent casualties. Maria followed Philip into the grave. And so the toddler's grandfather endured the always tragic reversal of the expected natural order whereby a child outlives his parent.

The old ship merchant in New York had no intention of losing his namesake grandson as well as his son and daughter-in-law. Mindful of medical theories that humid tropical vapors were to blame for epidemics, Frederick I shipped the boy back to New York without delay. He also summarily disposed of Spring Head, pocketing a cool ten thousand pounds for the property, a sum equivalent to more than two million dollars in 2005. Barbados still warranted Philipse involvement—family ships would continue to enter its ports to trade, especially to barter Africans for sugar—but the island's climate now appeared too dangerous to risk habitation for any length of time. As for the upbringing of Frederick II, middle-aged Catherine would become the child's adoptive mother and grandfather Frederick his adoptive father. Their ward would share their New York City home and their Philipsburg estate. Catherine and Frederick would raise Margaret's grandson with every luxury he could need or want.

Two years later, when Frederick II was four years old, the fever that took his parents seized New York City. A historic contagion claimed the life of Frederick Philipse, scion and merchant. Now twice a widow, Catherine had both to organize her husband's burial (a major society event in which each detail must be micromanaged, especially given the recent blemish on the family's reputation) and care for her adopted boy in a virus-ravaged environment the likes of which New Yorkers had never before experienced.

Yellow fever had been a far-off worry. It was supposed to spread in tropical climes, not the fresh North American air. Now, in three months, the disease killed 570 residents of the city, fully 12 percent of its population. The fever even snuffed the mayor of New York, while his administrative higher-ups fled to their country places to escape infection.

Looking for some plausible cause, much as they had when confronting comet sightings and crop failures, city leaders blamed the public's "manifold sins, immorality and profaneness." Doctors attributed the fever to the effects of miasmic marsh air—according to long-standing scientific theory, air and water themselves could transmit life-threatening illness. No one dreamed that a specific breed of mosquito, *Aedes aegypti,* could have transported the virus on merchant ships from its West Indies home to North America. This insect's preferred breeding ground was not sewers or swamps but clean standing water, such as that cistern of pure rainwater in the tidy backyard of a prosperous dwelling. And unbeknownst to anyone, this particular breed of bug fed by day rather than night, so that the bustling wharves of New York with all those tasty humans going about their business represented a veritable banquet. The inhabitants of the city swatted mosquitoes in annoyance, but they never thought to save themselves by exterminating the insects en masse.

Catherine and other survivors lived with the haunting reverberation of the death carts that rumbled continually through the dusty streets with their loads of bodies. An Anglican missionary counted twenty deaths each day for the duration of the outbreak, which amounted to a lot of carts in the still-small area the city occupied at the foot of Manhattan. And the church bells tolled ceaselessly for the dead. Colonists were terrified of being surrounded by fever, perhaps the least understood of all physical symptoms. With no tools to quantify a patient's temperature or medicines to reduce it, people believed that fever was not just a symptom but "was tantamount to disease itself" with "death . . . assumed to be a likely consequence," as one medical scholar explained. New Yorkers saw the specter of this particular assassin close up in their homes, where the typically cramped living quarters forced a constant intimacy with family members in extremis.

Caregivers, in large part female family members such as Catherine, witnessed the odd, meandering progress of the disease. Two or three days after the onset of symptoms, Frederick would have entered a remission so complete that he thought he was cured. But a day after getting up to go back to his routine, he would have been struck down again, and this time the decline was fast and sure. It was all over in a week.

During that time, Catherine would have wrung out the wet cloths to ease her husband's burning sweats, mopped up floods of black vomit, comforted him during fits of agonized hiccuping, and watched his skin take on the ocher tinge that presaged the end. No amount of tender treatment, cold baths, or bark distilled in water—the aspirin of the day—could cure this "malignant distemper" once it attacked. It targeted infants and octogenarians with equal virulence, and every age in between.

When Frederick succumbed in 1702 the merchant was seventy-six years of age, and he and Catherine had been married a scant decade. It was a good thing Philipsburg Church was completed. Catherine fulfilled her husband's desire to be interred in its crypt. Frederick's last will and testament also distributed the family riches among his children in the new English fashion, with the scion's male heirs receiving the preponderance of his estate, rather than the Dutch practice of dividing all bequests equally among siblings.

Eva, as promised by Frederick "out of special love and affection" when she was a fatherless baby, was not forgotten when it came to doling out the Philipse largesse. He did not, however, adhere to his pledge to "keep said child as my own," to treat Eva as a joint and equal heir on the same par as the rest of his children, as strictly as he might have done. Frederick bequeathed to his stepdaughter—and, explicitly, to her husband Jacobus van Cortlandt, since women must now keep their property jointly with their husbands—the house they inhabited in New York, one other lot in the city, a mortgage for a stretch of wilderness north of Manhattan, and "one quarter of all ships, plate goods, etc.," the last a formidable mass of property. These gifts complemented the seventy-six-acre spread at the southern boundary of the Philipse lands that Frederick had sold the couple in 1699, a fertile former Indian agricultural hub that would be known in contemporary times as Van Cortlandt Park.

All told, it made a respectable inheritance, a comfortable inheritance for a person of her parentage. Still, it was fortunate that Eva had a husband with resources of his own to maintain their fast-growing household (at the time of Frederick's death she already had produced four of her five children).

Annetje's slice of the family fortune was significantly larger. In tandem with her husband, an English merchant named Philip French whom she had married in 1694, Annetje received the Manhattan house in which she lived, one city warehouse, and all the family's Bergen holdings—most of which probably had originated with Margaret—including a house, garden, fifteen acres of land, sixteen acres of meadow, two farms, and a plantation, the customary term for an extensive estate. Frederick also gave Annetje and Philip 990 acres of rich Ulster County land and a quarter of all ships and goods to complement Eva's.

Annetje's inheritance, too, could not be considered anything but generous. Yet it was still a daughter's portion when contrasted with the jackpot awarded each of the two male heirs to the Philipse fortune.

In his will, Frederick gave Adolph his reward for acting as dedicated merchant in training. Four houses and a warehouse on Manhattan went to Frederick's second born. So did all the "lands, tenements, and hereditaments" at the Upper Mills, that property's fourteen "negroes," and the mill boat named the *Unity.* So did one-half of a sawmill at the village of Mamaroneck in Westchester, and all the Philipse properties at Tappan Creek in New Jersey, as well as one-quarter of the shipping business. This was a scion's inheritance, geared for the person most prepared to immediately take the reins of the Philipse enterprise. But it was not the biggest chunk of Philipse stock to be divested in Frederick's will.

It might have come as some surprise to witnesses when the will was unsealed that its most amply endowed beneficiary was the testee's four-year-old grandson, as of yet unable to effectively take the reins of anything but a gently rocking hobbyhorse. At some point Frederick had determined that the son of his eldest son Philip should be dealt an inheritance even more massive than any his own children would see.

Perhaps it all came down to a name. As the old man's name-sake, the toddler was destined to carry on the proud Philipse tradition, at least in the eyes of Frederick I. Or perhaps the prize originated in the pain the grandfather still felt over losing his eldest child, possibly his favorite child, just two years earlier. Whatever the cause, the result was that when Frederick II, "my grand son, born in Barbadoes, ye only son of Philip, my eldest son, late deceased," came of age, he would receive: three houses, two warehouses, and various lots throughout Manhattan, plus the "lands, houses and hereditaments" in the lower part of Philipsburg, extending from the "bridge and toll" at the Wading Place north to the Nepperhan River, and encompassing "all those lands and meadows called the Jonckers Plantations, with all the houses, mills, orchards, etc. within the Patent" and beyond to the old Indian site of Wysquaqua Creek in Dobbs Ferry and east to the Bronx River. There were a few other items thrown in, including a slave "called Harry and his wife and child," two additional slaves named Peter and Wan, and a mill boat called *Joncker*. Just like his aunts, the child would acquire a quarter of all the family ships and plate goods. For Frederick, however, these enrichments would amount to the icing rather than the cake itself.

Catherine's bequest was so simple, it took only a few lines to set down. Frederick would give "my dear wife, Catharine Flipse" a flat fifty pounds per year. She was welcome to "continue to remain in the house I now live in," the Stone Street house. When the boy came of age, Catherine's Stone Street house would accrue to him as well.

Catherine could not have been surprised by these financial provisions. At the time of her marriage she had agreed to keep their fortunes separate, and she undoubtedly still had funds of her own.

She might have been more surprised by a provision that was not strictly financial. The couple's union may have been brief and her material endowment scant, but their connection, it turned out, would last far beyond Frederick's burial. An assignment stipulated by Frederick in his will would occupy his widow for the rest of her life. Catherine, he instructed in a somewhat garbled passage, should have "the Custody, Tuition and Guardianship of my Grandson Fredrick Flipse and of his Estate to his. . . . ?[*sic*] Untill he comes to the Age of one and twenty years who I desire may have the best Education and Learning these parts of the World will offer him, not doubting of her care in bringing him up after the best manner possibly shee can." Despite the language, the directions could not have been clearer. Nor could the subtext. Frederick may have skimped on her financial bequest, but he charged Catherine with his namesake's upbringing as though that responsibility was itself a treasure and a privilege, with the boy's guardianship amounting in effect to a material aspect of his estate.

Dutiful Catherine undoubtedly had less energy than she would have had years earlier, before nursing and burying two husbands and supervising a major construction project. Yet the sober fifty-year-old widow dived into tending the needy son of her husband's deceased son. Never a mother, she would now mother this boy, her stepgrandson, on her own. She would obtain for him the education her late husband insisted upon. She would tolerate a fidgety child hanging on her skirts when she was ready to simply rest her joints in a chair—even the unpillowed straight-backed chair that was universal in well-appointed eighteenth-century homes. Frederick left Catherine with little materially but a handsome wardrobe, a roof over her head, and a conservative annuity. She was still the lady of the manor. But this small child would be saluted by all as lord.

II

Not Doubting
of Her Care

By the time Catherine returned to Philipsburg, she had been gone from New York, from both the city and the Manor, for nine years.

Her decision to leave the province after Frederick's passing seems to have taken some time. The New York City census showed "Widow Phillips" as an inhabitant as of 1703, in a household that included her stepgrandson and an entourage of seven slaves. Catherine's husband had been dead for a year, and it would have been easy to maintain a comfortable life for herself and the child in her well-appointed home on Stone Street, with so many family members residing in helpful proximity.

Yellow fever notwithstanding, New York in general was an increasingly comfortable place to live. Thirty-five years after the start of British rule, the city's population had swelled from two to five thousand. Its area had nearly doubled, with almost as much cleared land above Wall Street, the elegant thoroughfare that remained when Het Cingle fell in 1699, as below it. Elite New

Yorkers clustered, as they always had, at the foot of the island. The long-held, ever-appreciating properties of the Philipses and their cronies lined the fashionable lanes whose names remained unchanged from decades past—Stone, Broad, Lower Broadway, Pearl—and the new straight streets bearing English names such as Queene and Crowne. Catherine still lived only a few doors down from the house where she was raised, keeping an eye on Frederick as he played in the same street where she had played. In 1707 a baptismal record at the Dutch Reformed Church cited "Catharina Van Kortlant wife of Vredrik Flipse" as sponsor of newborn Johannes van Imburg, whose parents presumably were her friends. That was the last public document attesting to her dwelling in New York for nine years.

Most likely Catherine was looking for the best education for young Frederick. And that could not be found in Manhattan. The Dutch primary schools of Catherine's youth already had experienced the crisis that would lead to their disappearance. In 1705, after the death of New York's revered schoolmaster Abraham Delanoy, the Anglican governor Edward Hyde, Viscount Cornbury, seized the opportunity to deny licenses for two new teachers hired by the Dutch Reformed Church. The policy forced schoolchildren to attend the increasing number of English schools in the city, where the number of English instructors outnumbered those of Dutch or French descent by thirty to five during the first three decades of the century. Cornbury established a city grammar school run by Anglicans; in 1708 it enrolled 119 boys, taught by five English schoolmasters.

Girls attended school under English rule, but their curriculum had changed considerably since the days of New Amsterdam. The schooling of a girl Frederick's age likely consisted of the "Dancing, Plain-Work, Flourishing, Imbroidery, and various sorts of Works" offered by one Mrs. Brownell, whose tutoring complemented the reading, writing, ciphering, "Merchants Accompts," Greek, and Latin her husband taught the boys. Affluent parents also paid for their daughters to attain a mastery of painting upon glass, a great craze at the time, or to craft artificial fruits and flowers, while their brothers studied navigation, bookkeeping, and surveying. The fruits-and-flowers female curriculum of early eighteenth-century New York goes a long way toward explaining the absence of ship merchants such as Margaret Hardenbroeck in the new century.

The city offered few advanced scholarly opportunities even for boys such as Frederick. In 1704 influential landowner and politician Lewis Morris proposed that "New York is the centre of English America and a fit place for a college." Higher education, however, would not reach New York until 1754. At the start of the 1700s, young men could leave home to attend Harvard College in Boston, but the school focused more on training the Puritan clergy than landed merchants. William and Mary, a fledgling college in Williamsburg, Virginia, also had received its charter in 1693. Yale, in New Haven, opened in 1701. Yet neither could boast the reputation of a more established institution.

Like those in England.

Frederick's dying wish for his grandson to obtain "the best Education and Learning" in "these parts of the world" appeared something of an oxymoron. How best to resolve the dilemma? Catherine chose the education that anybody in these parts would recognize as the best and swept Frederick II off to London. The details of his curriculum are not documented. Perhaps she took him to one of the innumerable grammar schools that catered to England's younger thoroughbreds in a timeless, timber-framed setting. She might have encouraged him to advance to the next stage and enroll at Oxford or Cambridge, where he could grapple with the seven liberal arts: grammar, rhetoric, logic, music, arithmetic, geometry, and astronomy. Back in America, Frederick would introduce himself as an attorney. He must have earned his law degree in England during those nine years or later, on a trip abroad of shorter duration, because there were no law schools yet in America.

One thing is for sure: By the time Frederick returned to New York in 1716, he had steeped himself in all things English—an experience that flavored every aspect of his mature life in America. Born in a British colony, schooled in British classrooms, exposed to British customs and style, he found comfort in the hierarchical condition of monarchy. With his vast manor, tenants, servants, and slaves, Frederick was the closest thing to European gentry on American soil. Even the Church of England, it turned out, suited him better than the faith of his forefathers.

New York Episcopalians attended handsome, high-steepled Trinity Church, which had been built in 1696 of velvety brown hewn stone at Broadway and Wall Street. The edifice commanded a formidable view of the Hudson River and opened its doors during a church-building boom on Manhattan. It was joined by a Quaker meetinghouse in 1703 and a Huguenot church in 1704. Within the next ten years the city would see brand-new meetinghouses for Presbyterians and Baptists, and a new Dutch Reformed church. North America's first synagogue opened in a private home off Broadway on Beaver Street in 1691 (worshippers would not get a free-standing temple until 1730). Pluralism had the official support of the Crown: The Act of Toleration, passed by Parliament in 1689, guaranteed the right of all to public worship, as did legislation adopted by the New York Assembly two years later. Trinity Church expanded twice before 1737 to accommodate a burgeoning congregation that consisted not only of English colonists but also Huguenot and Dutch recruits. The provincial government favored the Anglicans from the start, which isn't surprising considering that the Church of England was the official faith of the Crown.

But many Dutch colonists resented the preferential allotment of land and money to the English church.

Lord Cornbury had awarded the church a west-side tract known as King's Farm that stretched from Wall Street all the way north to what would become Christopher Street. He also persuaded the assembly to fund a generous raise for Trinity's rector and welcomed the Society for the Preservation of the Gospel, the Anglican missionary machine, to sign up recruits around the city. In 1707, despite the Act of Toleration, Hyde threw Presbyterian clergymen in jail and replaced them with Anglican ministers.

Most of the old-school Dutch despised the governor. It didn't help his reputation that he was said to indulge in public crossdressing. "His dressing himself in women's Cloths commonly [every] morning is so unaccountable," said Robert Livingston, "that if hundred[s] of Spectators didn't dayly see him, it would be incredible." The governor's wife, for her part, gained a reputation as a stickyfingered party guest who pilfered cutlery from the homes of genteel hostesses.

Parliament relieved Governor Hyde in 1708, and the New York merchants to whom he owed money sent him to debtor's prison. Yet the power of the Episcopal Church did not abate. It did not help that many Dutch Reformed parishioners began to leave their church, partly in disgust with their own clergymen, whom many thought did not stand firm enough against the Anglican presence. (And in fact the Calvinist and Episcopalian leaders were more interested in bridging their differences, finding common ground between the two Protestant faiths, and forming a united front against the despised Catholics.)

Increasing intermarriage between Dutch and English colonists resulted in a watering-down of the Dutch Reformed observance. Language figured in this evolution. With Dutch schools on the decline, a generation of children grew up with little knowledge of their ancestral tongue. They wanted church services conducted in

English, in a language they could understand. By the turn of the century, a visiting Boston doctor named Benjamin Bullivant chronicled a widespread disinterest in the Sabbath among Manhattan's Dutch residents: "some shelling peas at theyr doors children playing at theyr usuall games in the streets & ye taverns filled." Peter Kalm offered his analysis in 1749, writing that in New York, "most of the young people now speak principally 'English' and go to the 'English' church; and would even take it amiss if they were called 'Dutchmen' and not 'Englishmen.'" The Church of England increasingly appealed to a generation ready to abandon its Dutch roots.

Like many of his privileged peers, Frederick Philipse II knew the politic place to be on a Sunday morning in Manhattan. Later, he would ensure that the Anglican Society for the Propagation of the Gospel found a warm welcome in Westchester. And later still, he would secure the future of the Church of England in Philipsburg by releasing the capital to erect a grand Anglican church on the Nepperhan, just three hundred yards down the road from Margaret's Manor Hall. As he grew older, Frederick almost surely regarded Catherine's church, the church of his ancestors, as ascetic, dusty, and irrelevant. He found the powerful Church of England, with all its gaudy ceremonial trappings and royal support, a more comfortable fit.

Early on Frederick showed an appetite for power. It went beyond religious observance. He had no interest in building things with his hands, as his grandfather Frederick had done as a young carpenter. He would not engage in the trade of basic necessities, as his grandmother Margaret had done throughout her life. Nor would he attempt to open new markets, as his father, grandparents, and uncle Adolph had when they embraced the slave trade. He did not want to be a merchant like the rest of his clan. And he did not have to be one, thanks to the resources his family had given him.

Frederick wanted to be a gentleman. In his era, gentleman was synonymous with political operative. The role was a luxury, attendant upon a fortune adequate to allow a man to spend time maneuvering, attending council meetings, and socializing with cronies. In 1719, the year he turned twenty-one, Frederick gained the position of alderman for New York's South Ward, the Philipse family's home turf and one of the city's six jurisdictions. His youth proved no disqualification for so significant a political placement. After all, it was counterbalanced with his family name, not to mention his net worth, which along with support from the political powers-that-be defined a man's eligibility for office.

Frederick had come of age. Officially acquiring all he had been willed—his various homes and farms, his mills and mill boat, his slaves, and his ships and "hereditaments"—Frederick became one of the wealthiest men in all of New York. His portfolio had appreciated under Uncle Adolph's astute management, but Adolph's advice was not particularly necessary any longer. Catherine's role as caregiver officially expired, too.

Catherine found herself with stretches of time the likes of which she barely remembered. She could stay at Stone Street, if she liked, or ride north to Philipsburg. She could overnight in the house on the Nepperhan if she wished, if Frederick did not need it.

The country had its advantages. In October Catherine liked to wander the long blond meadow grass between the trees to select perfect apples—the Newtown Pippins with the firmest, brightest skin or the largest ripe greenings for baking. When she took the Post Road from the Manor Hall to her church on the Pocantico, she passed at least a dozen fruit orchards snugged up against small houses like the one Margaret had built, solid yet modest efforts in rubble-stone or wood.

She also passed fields of grain, some of them growing straight up to a farmer's doorstep and enclosed by worm fences, enclosures whose name derived from their sinuous skeletons of branches. A few taverns dotted the landscape, each with its roaring fire on the ground floor and a few creaky rope-strung beds upstairs for people going a distance. Most also offered pastures out back for northern drovers to park their cattle overnight before heading south to the meat market in Manhattan. The taverns drew together the still-isolated tenants; the census of 1712 showed a population of only 608 for the entire manor. The Philipse lands were dotted with farmsteads, and each year resident families traded sons and daughters to combine families until the generations shared given names, surnames, and physical characteristics.

The Nepperhan had more mills now, but there was still an aura of unbroken peace, and it seemed the loudest sounds came from the overhead rustle of the first-generation trees and the tumble of water falling toward the Hudson. The big river supplied interesting sights and frequent visitors. Most downstream traffic originated in Albany, as it had in Margaret's day. A stream of pelt-laden canoes descended from the Mohawk River, and fleets of sloops, smacks, and the fast sharp-ended boats known as pinks glided by the bank of the Hudson below the Manor Hall as if for Catherine's private entertainment.

In the Hudson Valley, trade remained the linchpin of the economy, but the product had changed. Furs no longer were golden. Lumber commanded top prices, though the cherished forests of Philipsburg stood barely touched. The family's tenant agreements spelled out an unswerving restriction barring the use of timber by anyone but the lord himself. As for the slave trade, Adolph had taken up his father's mantle, but not until the second decade of the eighteenth century, once the burn of the Madagascar experience had healed. Then Adolph returned to Africa aggressively, expanding his investment and bringing home hundreds of captives rather than dozens. Adolph became not a dabbler but a player in the human-trafficking trade, which brought an estimated sixty-eight hundred Africans to New York between 1700 and 1774.

The business partnership between Adolph and Frederick, however, increasingly revolved around a less controversial product: wheat. Flour milling had grown into a booming concern, with the triumvirate of flour, biscuit, and meal exports second only to tobacco. From the Philipsburg Manor wharves at the Pocantico and Nepperhan rivers, sloops carried grain harvested by one hundred tenant farmers to New York City for sale abroad and in other colonies. Years earlier, before Frederick came of age, Adolph had joined with Philip French, his sister Annetje's husband, to oversee the manor's land, mills, smithies, coopers, and tenant farmers. (With Annetje's husband hailing from upper-crust Kelshall, in Suffolk, England, and with the sharply divided marital roles of English custom increasingly the norm in New York, there is no reason to think Annetje took an active role in managing the commercial affairs of her family. She had a household in Manhattan to occupy her energies and four children to raise.)

The phenomenal value of grain-related products is hard to appreciate today. How could someone make a fortune—not a living, a fortune—from so simple an agricultural product? Not to mention from baking and selling the modest component of every ship's larder known as biscuit, an unadorned cracker with the taste of cardboard? In part the explosion of the family's wheat fortune derived from the totality of the enterprise. Philipsburg was a one-stop shop for grain, an agricultural staple with a seemingly bottomless market both overseas and domestically. The Philipsburg operation had every step of the process down to a science: from growing the grain, to producing flour and tack, to packaging and transporting the products.

The Philipse family gristmills were mechanical marvels constructed almost entirely of hand-carved wood and stone. Each mill was powered by a gear shaft—one mammoth log of oak, twenty-six feet long and weighing nine thousand pounds—running from the waterwheel to the compact intricacy of wheels beneath the mill floor. The Hudson Valley had imperfect soil, but a perfect temperate climate and the water necessary to power the mills to grind the wheat and transport its products for sale to other markets, mostly across the ocean.

Wheat flour made for an economy rooted not in the abstract currency of paper money but in a rich substance an investor could actually touch. If you held your fingers under the mill chute, you could cup in your hands the creamy, buff-toned flour as it poured out, dense with the oil of the wheat germ and warm as breath, with the sweet smell of earth. Two sets of millstones pounded the pink-centered wheat berries between them, a grind set in motion by the power of the rushing waterfall over the milldam. The stones, European granite imported from Holland, lasted half a man's lifetime. It took a team of draft horses to drag just one up from the wharf. Each was the size of a truck tire, with a flat surface arduously dressed in specially patterned grooves; a master craftsman had worked to create a precise scissoring effect with the aim of grinding the wheat, not pulverizing it. Rough meal fell through the center hole of the lower stone, the bedstone, then was taken up to be ground over and over again until its consistency was perfect for making dough. A Philipsburg mill could churn out fifteen tons of flour in an average week.

Mill workers sieved the flour through silk to separate out its dark bran, and master coopers packed the product into white pine barrels constructed of precisely contoured staves. The cooks in the bake house kept the ovens heated year-round to transform the dough of water, flour, and salt into biscuit, also known as hardtack. Before long new gristmills clustered around the first, along with sawmills, coopers, and smithies. Taverns opened. While wheat and other grains formed the foundation of both the diet and commerce of New Netherland, mills also stood at the society's center socially.

Slaves were in charge of each Philipse mill operation. At the Upper Mills, on the Pocantico, a Dutch overseer named Aert Willemse stopped in regularly to attend to the books; he was the only regular nonslave presence on the estate. At the Lower Mills, in Yonkers, another group of enslaved men and women operated a comparable grain-processing operation. Frederick I had left Adolph with fifteen slaves when he died in 1702, a cadre Adolph supplemented over the years with labor brought in his own ships from Africa or the West Indies.

The black residents of Philipsburg maintained a virtually independent colony quite different from the common clichés of plantation labor. Owner and overseer were largely absent from the premises. Many slaves became specialists, masters of one of the trades the manor depended on, and so the complex thrived with its constellation of wheat farm, cooper shop, bake house, and dairy. Butter, heavily salted and packed tightly in nine-gallon barrels called firkins, was a staple of the West Indian trade. Philipsburg female slaves perfected its production.

Whether at the Upper or Lower Mills, the estate's men and women proved to be capable artisans and mariners, and they provisioned themselves with the garden plots they were assigned, harvesting root crops, sweet potatoes, squash, greens, and especially kidney beans, probably their most important staple. The slaves even raised surplus produce to sell at a weekly market in West Chester, a village miles south near the bottom of the mainland, east of Spuyten Duyvil. They traveled there alone to hawk their cabbages, parsnips, and radishes, indicating a degree of trust between the holder and his slaves, or at least a resignation to the status quo.

An ad placed in the *New York Weekly Journal* offering a slave for sale suggested something else—a certain level of appreciation and even respect for the abilities of a twenty-year-old woman who could do "all sorts of House work; she can Brew, Bake, boyle soaft Soap, Wash, Iron & Starch; and is a good darey Women she can Card and Spin at the great Wheel, Cotton, Lennen and Wollen, she has another good Property she neither drinks Rum nor smoaks Tobacco, and she is a strong hale healthy Wench. She can Cook pretty well for Rost and Boyld." Her only defect, according to the seller, was her inability to speak any language besides English, but that was perhaps offset by the fact that she already had the small-pox in Barbados as a child and so was immune.

The slaves of New York possessed some advantages compared with the southern plantation slaves, almost all of whom performed backbreaking unskilled labor and slept in hovels far removed from the comforts of the Big House. Northern slaveholders depended upon their chattel for skilled work more than the drudgery of field-work. Slaves were not discouraged from educating themselves in reading, writing, or any useful craft. And owners valued some in-dependence on the part of slaves, especially in the agrarian lands sur-rounding New York. At Philipsburg, enslaved men and women cared for the place as if it were their own.

That did not mean New York slaves were satisfied with their lot. At the start of the eighteenth century the black population on Manhattan, the greatest concentration in the province, numbered about two thousand, or roughly one-fifth of the city's inhabitants. The population constantly expanded as newcomers arrived at the slave market at the foot of Wall Street that opened for business in 1709. No matter what their skin color, virtually all city dwellers resided within five square miles at the island's foot, south of the fu-ture Washington Square. The races occupied extremely close quar-ters; enslaved blacks cohabited with their white owners in the finest neighborhoods.

A strong African flavor pervaded the town. The number of enslaved people in New York, in fact, exceeded that of any other

English settlement aside from Charleston, South Carolina, and the proportion of blacks to whites was higher than in any other northern settlement. (The southern rice-indigo-tobacco-growing colonies, of course, had the highest ratio of all, with fully one-third enslaved men and women.) The system of slavery was totally entrenched in New York, with at least one slave in 40 percent of city households.

Racial tensions had mounted since the start of the century. In 1712 two dozen well-armed slaves from the Coromantee and Pawpaw tribes of Africa gathered at midnight in an orchard on Manhattan's Maiden Lane. After setting a nearby shed on fire the slaves, armed with hatchets, knives, and swords, ambushed the white men who arrived to douse the flames. Under pressure to explain the tragedy to higher-ups, Governor Robert Hunter offered this disingenuous interpretation of the rebels' motivation: "They had resolved to revenge themselves, for some hard usage they apprehended to have received from their masters (for I can find no other cause)."

Twenty-one slaves were slowly roasted, hanged in chains, or broken on the wheel after the attack; six slit their own throats rather than submit. Two slaves by the names of Abigail and Sarah were among those convicted as accessories in the rebellion. Both died by hanging—although the execution was delayed until after one of the slaves delivered her baby. Even an unborn child constituted the valuable chattel of its owner.

The executions sent a strong message, yet the city's elite could not slake the sense that this would not be the last rebellion. To avert any further unrest, legislators expanded and enforced a slave code that was already on the books, reversing a long-standing flexibility in the treatment of African men and women. As of 1712, New York's slaves could not travel independently past dusk (a particular inconvenience for the Philipsburg mill operation, which customarily sent its master boatmen down to New York Harbor with a load of product). Slaves could earn forty lashes just for assembling in a group of more than three. And black New Yorkers were barred from owning property, a provision that demolished rules in place since the colony's founding by the Dutch.

Despite the new ordinances, many slaves managed to escape their masters, a phenomenon documented by hundreds of newspaper advertisements seeking runaways. Nearly every notice included a vivid compendium of attributes, from the color, fabric, and style of the runaway's garments, to his physical traits, affect, disposition, and mastery of languages and skills. The ads demonstrate why affluent New Yorkers, those who could easily afford to buy a fresh African at the slave markets, would be sufficiently dependent upon these particular men and women to chase them down.

A runaway named Clause, for example, cited in a 1730 ad in the *New-York Gazette,* was about twenty-seven years old, wore an old striped vest, leather breeches, and black shoes with buckles. He spoke both English and Dutch and played the fiddle. Fiddlers abounded among the slave community. Of 753 notices published between 1716 and 1783, 6 percent (forty-four ads) referred to the runaway's proficiency on the fiddle, while a smaller number were identified as drummers, fifers, or French horn players. Other skills cited include farmer, unsurprisingly, as well as weaver, barber, doctor, and even magician. The ads hint at epic tales: Four men took off in a two-masted cedar boat and pine canoe, packing a "white flower'd serge jacket," "three Kersey Jackets and breeches of a dark colour" and "a green ratteen Jacket almost new," along with cheese, bread, and a firkin of butter. One of the men, named Venture, was described as six-foot-two and "large bon'd," "scar'd with a knife in his own country," while Isaac had a "bushy Head of Hair" and a "sower Countenance." Their master offered twenty pounds to whomever could secure the runaways.

Other slaves could read and write. According to a doctor who placed an ad, his runaway slave "speaks good English, understands all Sorts of Farmer's Work, and something of the Sea, and no Doubt will endeavour to pass for a free Negroe as he can write any Pass he thinks necessary." A "mulatto wench" named Fanny with "a smiling countenance and black-curled Hair, full-breasted," fled wearing a small black silk hat, a blue and white striped gown, and a red petticoat. Fanny was clearly a force to be reckoned with, and her master, a Mr. Thomas Scramshaire of New York, seemed to have committed to memory every facet of her form.

Over the years, the Philipse family saw its share of runaways. In 1693 a slave named Jack—described by Frederick the elder as "a remarkable fellow, looks Extreem Squint, Speaks verry good English & Dutch, and is of Statture verry Tall"—escaped from Philipsburg Manor. Convinced the man had reached the island of Sainte-Marie, Frederick told his ship captains headed to Madagascar to watch out for this "remarkable" slave and to fetch him back to New York. Jack, however, never resurfaced.

Frederick and Margaret's son Philip also lost a slave, a sailor named Nick who served as an expert "linguistor," or interpreter, for the Philipses' Madagascar business. In the 1690s, when Nick arrived in Sainte-Marie on the *Charles*, Adam Baldridge managed to purchase the man from the master of the ship. Frederick sent the *Katharyn* to Sainte-Marie with instructions to buy Nick back. But Nick negotiated, and used funds he had earned assisting Baldridge in the slave trade to procure his liberty from Frederick, who signed an official document that pronounced Nick "manumitted, Sett at Liberty, quitted and discharged." Frederick must have decided it was cost-efficient to be done chasing this particularly stubborn piece of property, as valuable as he was.

The individual value of enslaved men and women was a significant factor in the development of Philipsburg Manor. The men and women who transformed seeds of grain into the commodity that enriched the Philipse family fortune were highly skilled, highly disciplined, and dedicated to their craft. The enslaved men who acted as millers, boatmen, or coopers had slaves who worked for them, reporting to them in the same way any enterprise employs leaders and apprentices. These master craftsmen held authority. They had to believe in the importance of their efforts to produce the high-quality grain, flour, and biscuit that mills at Philipsburg loaded up for export. And in exchange for the quality of their labor, they earned a measure of respect from all tenants who brought their grain to be ground, packed, and shipped from the Philipsburg mill complex. It wasn't freedom, so it had its own flavor of bitterness, but it was something.

$\mathcal{N}$one of Philipsburg's bounty actually belonged to Catherine, of course.

Frederick II enjoyed legal ownership of the house in which Catherine had lived off and on over the years, the house his grandmother Margaret had built on the bank of the Nepperhan. The Manor Hall's rough stones were his now, as were the honey-toned floorboards, the cavernous hearth, the view of ancient cliffs on the west side of the Hudson. Even as Frederick matured, Catherine had never worried about being displaced; she was essentially the boy's shadow, with the all-important job of chaperoning and protecting the young lord. Frictions naturally developed between them, differences over their values, Catherine's old-fashioned ways, her old-fashioned faith, and young Frederick's embrace of all things English. But those differences could be chalked up to a generational divide. They had always shared a compact, an understanding that Catherine would nurture Frederick and he would shelter her.

Then Frederick found another woman.

Joanna Brockholst had also spent her formative years in a condition of uncommon privilege, born to a family of prominence and power comparable to that of Frederick's grandparents. Her father,

Anthony, who had immigrated from Amsterdam, served as Crown-appointed lieutenant governor and then governor of New York in the 1680s. Throughout a distinguished career he made himself useful to the British administration, both as military officer and as political functionary.

In 1697 Anthony Brockholst relocated his family to an expanse of land in northern New Jersey known as the Pompton Valley, after the Indians that made their home there (the name survives as Pompton Lakes). Europeans were making their first forays into obtaining Lenape land, and Brockholst and six other prominent men purchased five thousand acres for "wampum and other goods and merchandise to the value of 250 pounds current money of New York." Born in 1700, Joanna was the next to youngest of the Brockholst children. Watching her parents manage the sweeping environs of their Pompton property figured as an integral part of her formative years.

Joanna and Frederick shared a taste for English ways that seemed ordained by the upbringing of each—Joanna in the orbit of a powerful father with close connections to the new imperial culture of New York, and Frederick from his London acculturation. By the time they married in 1719, the colony had grown significantly less Dutch. Pockets of the old, organic Dutch culture persisted, particularly in the Hudson Valley and Albany area. But the population was irrefutably shifting. In the 1670s 80 percent of white New Yorkers were Dutch. The proportion was steadily dropping, on its way to 40 percent by the 1730s.

Theirs would be a traditional aristocratic English marriage, with plenty of babies. They did not wait a year to have their first—Frederick III arrived in 1720—and once they got rolling they were unstoppable. When it came to introducing those babies to the world, the couple made a ritual tribute to their Calvinist ancestors: Records of New York's Dutch Reformed Church show that the

Philipses brought in eight of their offspring for baptismal rites. (Three infants did not live to receive a baptism, and the certificate for one, Mary, has disappeared.)

Now Catherine, dowager-of-all-trades, lived everywhere and nowhere. She attended at least one baptism of Frederick and Joanna's babies, according to church records, suggesting that she played at least an honorary role in the life of the couple. Yet her mission of raising an educated gentleman had been completed. She kept a bed at Adolph's, a stone's throw from the church, whose parishioners noted her as strong, even somewhat steely, but generous and fair. She had a bed at the Manor Hall, and squeezed in with Joanna and Frederick when they were there.

Catherine's house in Manhattan was just down the street from the family home she grew up in and it was her preferred spot now. She rattled around, alone except for her servants, her boy gone. She had lived past nearly everyone in her family—Stephanus and her parents in the grave, Maria also long gone, Jacobus still healthy but preoccupied with his wife and children—and survived the epidemics and hardships of the age like the tough old bird she was.

As the colony grew increasingly anglicized, the women of the province still displayed the strengths that startled visitors to Holland centuries before, but they had become less autonomous. Catherine was independent enough to build a grand church and carry her step-grandson off for the requisite gentleman's education, but she would never presume to represent herself in court, let alone stand for her husband. She traveled across the ocean to England, a voyage that took no less fortitude at the start of the eighteenth century than it did when settlers first landed at Manhattan. Nonetheless, Catherine sailed as a passenger, responsible for herself (and her boy), rather than as factor, in charge of the extensive cargo on board the ship. If Catherine owned ships, houses, or plots of land, no record of her ownership survived.

Catherine had built a life on ministering to others. And when she fixed her plans at the end of her life, at the age of seventy-eight, in 1730, she made a choice that probably never would have crossed Margaret's mind. Her last will and testament, witnessed by Frederick II when Catherine was "sick & weak in Body but of sound and perfect Memory and Understanding," represented her final ministration.

Both the substance and style of this document departed considerably from the highly detailed, appurtenance-rich testament of her deceased husband Frederick I. Being a woman of devout character, she began with an earnest digression: "first & principally I commit my precious Soul into the merciful Hands of Almighty God who gave it hoping for pardon and Remission of all my Sins in & through the Merits & Intercession of My Blessed Savior and Redemer Jesus Christ." In a similar vein, Catherine's principal bequest concerned an oversize silver beaker "on which my name is Engraven" that she left for sacramental use by the congregation at Philipsburg Church, along with a damask tablecloth "five Dutch ells" in length and a long table, with the stipulation that all three sacred objects be stored at Adolph's "Mansion House" on the Pocantico for safekeeping. (Philipsburg's longtime dominie Guilliam Bertholf had predeceased Catherine, and perhaps she did not yet trust the new reverend to safeguard her gifts.)

She then turned to the distribution of her valuables, the goods that belonged to her alone. It is impossible to say how much the woman was worth at the close of her life, because her will did not spell out value in real numbers but in percentages of "my whole Estate Both real & personal." Even so, the good Widow Philipse appeared to have spared no trouble in delineating a fair structure for her largesse. It is difficult to imagine that any one of her beneficiaries received a portion that was anything other than token, because her estate was split equally between the heirs of her five brothers and sisters (all but one, Jacobus, now deceased), so that each part was further divided into five or more parts. She did not leave any of her holdings to her Philipse stepchildren, who after all had plenty of property already. She also disregarded the British custom of bequeathing goods exclusively to male heirs.

As for Frederick II, though he served as witness to the will, he was not included among Catherine's beneficiaries. The fifth of a fifth share of the estate that emerged as Catherine's preferred gift was obviously something that she judged as unnecessary for the lord of the manor.

There was one final bequest outlined in Catherine's will, a stipulation that was not monetary in nature. Though it involved no transfer of goods, and enriched no one, this was the gift that perhaps expressed more than any other the widow's innate nature. With her dying words, Catherine directed that her personal servants, "Molly and Sarah My Indian or Mustee Slaves," would "be manumitted and sett at full freedom and Liberty." Their emancipation would commence "one Month after my decease."

Sarah and Molly: These were the women whose everyday existence had revolved around waiting upon their mistress in her chamber, carrying her toasty foot stove to church while they shivered in the unheated upper gallery, spooning her pap when she could not manage solid food, and emptying her chamber pot. Yet theirs was a physical intimacy surely matched by a certain mental closeness. Molly and Sarah likely slept on a pallet on the floor of the kitchen or outside Catherine's door, or even beside her bed, as was customary for Hudson Valley slaves. The three had heard one another breathe in the night numberless times when no one else was listening.

From over the centuries, it is hard to pinpoint one motivation for Catherine's act. Perhaps Catherine ended up caring so deeply about Sarah and Molly that she could not tolerate the thought of their possible mistreatment at the hands of a future owner. That was one explanation other dying slave owners gave when they emancipated their servants. A tradition existed, too, of freeing an individual who demonstrated outstanding loyalty during the years of an owner's wasting illness. Perhaps in this case it was just as much a way of honoring all that this nurturer had received from the black women and men who were central to her existence over the years. Enslaved Africans had hauled the stones and dug the mud for Catherine's beloved church. The trade in slaves had bankrolled Frederick II's law education. The servants of her household had pressed the petticoats, toted the Bibles, and deployed the complexion masks with which the aging lady of the manor cosseted herself.

Now that Catherine was dead, Molly and Sarah would be free.

But perhaps not.

A law already on the books when Catherine died stipulated that any owner wishing to liberate a slave must pay the government two hundred pounds sterling as security. With seemingly no knowledge that the manumission of her servants might be conditional, that it would depend upon any factor other than her desire to free them, Catherine apparently did not set aside the four hundred pounds required, or indeed any funds whatsoever for that purpose. She might have fallen out of touch with current developments in the province. Perhaps she did not possess the funds. Perhaps she hoped that her caring alone would make it happen.

When the old woman passed away, it was no doubt with the self-assured conviction that she was doing the right thing. And that, for a privileged, cash-poor New York woman with all the sense of duty in the world, was probably the best she could do.

12

Fashion Babies

The marzipan hedgehog lolled in a small slick of clarified butter, flanked by three baby hedgehogs with blank raisin-eyed stares, its plump, nut-paste body spiked with quills of slivered almonds. Its bed was an oval platter of blue-figured china, and a chunky pewter serving spoon lay alongside, ready to scoop the sweets onto a neighboring stack of diminutive plates. Joanna Philipse, the third mistress and second lady of the manor and its doyenne of lavish desserts, had spent two hours in the kitchen with her daughters effecting this masterpiece of whimsy. She gave Margaret, Mary, and Susannah the chore of crushing two pounds of blanched almonds with a pestle in a marble mortar, and glanced over every once in a while to see that they were doing the job without dropping the pestle on someone's toe. Then Joanna and the girls combined the ground nuts with cream, sugar, rose water, and egg yolks so bright they turned the batter saffron. They stirred the pot over the fire, sweat pouring from each cook's brow, until the ingredients congealed into a stiff candy clay. Children, even older girls, liked to assemble the mama hedgehog and her young because they were adorable; a socialite like Joanna served the treat because it was witty—hedgehogs, of course, were native to the English countryside and nonexistent in New York's Hudson Valley. And guests at a dessert spread had a keen appreciation of confectionary wit.

Churning out the requisite dishes for a party was demanding, especially in an inadequate kitchen. Since she first married Frederick II as a nineteen-year-old ingenue, Joanna had known the Manor needed upgrading. She set out to expand the wooden lean-to that served as the Manor Hall's kitchen after winning the grudging approval of her new husband's stepgrandmother, Catherine, who surveyed their doings like a stately buzzard. Joanna did not change it much, just pushed back the walls a bit to encompass a greater work and living area, since every kitchen at the start of the eighteenth century represented both. Quarters still were close, though, and got closer as the children started to arrive, young Frederick the year after the wedding, Susannah in 1723, and Philip in 1724.

Finally, early in the marriage, Joanna and Frederick decided that they must have more space in Yonkers. Their approach was basic: They doubled the area of the house on the Nepperhan, adding a southeast parlor to the right of the front entrance, and its upstairs counterpart. The addition was perfectly symmetrical, as was architecturally apropos for the Anglo-Dutch vernacular of the time, down to the placement and size of the added windows. The young couple made no drastic changes in form, used no new materials, despite the fact that the stonework of the original structure was clearly out of date. Style seemed of less importance than expedience, or perhaps a lingering respect for the ways of an early generation. Most important was the change in size. The house was still a box, but now it was a bigger, four-room box, commodious enough to house the burgeoning Philipses when they came to the country for a warm-weather visit.

For years thereafter, Joanna found herself too busy to consider any further renovations of the house. Instead, she delivered infant after infant, a total of twelve before she reached the age of forty-two. Seven of her children went to their graves early, but not before their parents had bestowed upon them cherished family names: Maria, Anna, Catherine, Eva, Anthony, Joanna, and Adolphus. The two girls after Susannah who survived their babyhood were Mary and Margaret, named for Frederick's legendary grandmother. With the five children in tow, as well as the omnipresent Catherine, Joanna made her way about New York, the busy wife of an up-and-coming politician. She absorbed herself in the ancestral home on Stone Street. It was not until the late 1730s that she began seriously to rethink the house on the Nepperhan as a place her brood could reside rather than just visit. And reside in style.

From the time he was twenty-one, Frederick had pursued a career trajectory that typified a man of his class and privilege. His grandfather, Frederick I, one of the richest merchants in the province, also had been a man of politics. He had sat on the governor's council for twenty-five years (until he was bounced as a pirate abettor, a subject best not discussed in the company of his descendant-heir) and before that he had held positions as city surveyor and as alderman. The elder Frederick had served as impartial arbiter in many delicate court cases. On one occasion, in May 1691, when Governor Henry Sloughter had to leave town for urgent Indian talks in Albany, he appointed Frederick Philipse co-commander in chief of New York City.

Frederick I was not the only member of his generation to build his politicking upon a separate career. Joanna's father, Anthony Brockholst, who served at one time with her grandfather-in-law Philipse on the governor's council, first made his name as a military officer. In 1677, at the direction of Governor Andros, he landed one hundred English soldiers and a seven-gun prefabricated redoubt at Pemaquid, Maine, to defend this Crown territory from insurgent Indians, the French, and pirates. For some time afterward England designated Pemaquid the port of entry of all of New England, and it gained greater prominence than Boston. Brockholst's heroism earned the career soldier his later political placements. (It was he who received the Albany townspeople's pleas during the comet crisis of 1680.)

Frederick II differed from his forebearers in at least one respect. He came to his first governmental posts with a résumé that was virtually naked. His gentleman's education as a lawyer—and his family pedigree—represented the full extent of his qualifications for public office. Nor would he ever undertake a profession to complement his political career. He did not want to, and he did not have to. When Frederick entered his first electoral race in 1719 to become alderman for the South Ward of Manhattan—representing what was still the district of the city's premier households—he was fortunate to have a relative with considerable clout to vouch for him. His aunt Eva's husband, Jacobus van Cortlandt, a powerful ship merchant and landowner, already had served one term as mayor of New York, in 1710–11, and was on the ballot again in 1719. While it is impossible to place a precise value on the grease a Van Cortlandt's support could provide, Frederick must have felt at least a little grateful to his uncle on election day, when the mayor-elect and the new aldermen went "in their formalities" from City Hall at the corner of Wall and Nassau to "His Majesty's Garrison Fort George" to take their oaths of office.

That was the first and only pull Frederick II needed. During the fourteen years that followed, the second lord of the manor won continual reelection as alderman. He also served as justice of the peace and acted on a variety of committees that dealt with such issues as maintaining the city's docks and bridges, surveying city lands, and establishing New York's very first "Publick Library,"

with a collection of 1,640 volumes donated by "the Venerable Society for Propagating the Gospel in Foreign Parts." He also held
municipal posts in Westchester, serving on the county highway
commission and as a member of the New York Assembly. But his
most distinguished office was Supreme Court justice for the colony. Frederick ascended to the bench in 1731 when he was just
thirty-three-years old and remained a justice for the rest of his
life.

Like most public officials, Frederick Philipse II had his share of
detractors—longtime council member and man of letters Cadwallader Colden condemned the judge as being "without pretence to
any skill or knowledge in the law"—but he did not let criticism
derail his sense of duty. Nor would Joanna Philipse wilt on those
periodic occasions when she saw her husband's judgment under
fire, in public discussion or in the newspapers, which had been in
broad circulation in the province since 1719. Joanna fathomed with
the cool sensibility of a political wife that any doubts about the
qualifications or performance of a civic leader could be considerably offset by inviting his peers to successful parties in his home.

*I*n 1745 Joanna and Frederick completed the transformation of
the house on the Nepperhan into a truly grand residence that
begged for parties. Already, with its four rooms, full attic, tile-
paved cellar, and wood-frame kitchen, the family home trumpeted
Joanna and Frederick's ever-expanding net worth. Some of the new
elements they had installed would be admired, even envied, by
their peers. They had constructed the fireplace in the new, southeast parlor in the English mode rather than the Dutch. Instead of
the delft-tiled jambless monster of Margaret's *groot kamer,* they built
a smaller, contained hearth befitting the family's dignified modern
style. The furniture stood square against the walls to make more of

the space (the room's dimensions were not that large, after all) but they were fine contemporary pieces, some of them French.

Then, like their wealthy friends, Joanna and Frederick II were able to afford new technologies that provided an extra draft of sunshine by day and more artificial illumination after the sun had set. The importance of these two features could not be overstated at a time when the majority of colonists went to bed with the sun rather than break out a precious wax candle.

One of their first renovations called for significantly enlarging the windows, transforming meager panes set in leaded casement to four-by-three sash-weighted, double-hung specials. Their sheer number made a statement: Even at the turn of the following century, the median number of windows for American houses was a paltry three. Houses in the countryside still largely relied upon an open door for daytime illumination. All the windows were framed by heavy drapes that puddled on the understated finish of the pine boards, yet another advertisement of the Philipse affluence. Few home decorators could afford to spill that extra dollop of pricey textile on the floor. Each downstairs room had a corner prospect.

Then Joanna and Frederick went beyond window treatments, with a more dramatic renovation that nearly quadrupled the size of the Manor Hall. Early in the 1740s, workers erected a spanking new wing that jutted north from the original structure so that the house took the shape of a backwards L with its foot, the old wing, extending toward the Hudson. The new wing had two stories, plus a full, finished attic and a sizable dry basement. Downstairs, entering by the east-facing new main entrance, guests faced a formidable curving staircase and a railing of polished, turned mahogany. This north wing held a comfortable dining room with its own fireplace. New bedrooms lined the floor above. The new wing was sheathed in smooth red American brick, although the family left the gray stone of the south facade and even matched its granite on the new wing's north end, as if to pay homage to the rough yet honorable ways of Philipses past. (The continuing attraction of stone is clear in another house. Eva van Cortlandt's son Frederick—Frederick II's cousin—erected a stone-fronted mansion in the Bronx in 1748 that is a virtual homage to Philipse Manor Hall.)

With the creation of the east wing, the west wing was also transformed. The southwest parlor, Margaret's *groot kamer,* formerly considered a chamber as elegant as any in the house, had now been reenvisioned as a business office for the estate. The ancient south-facing front door—the giant that Margaret had installed to make a showpiece entryway—became a service entrance. The change had a pragmatic element: tenants coming in with their dirty hands would not distress the lord and lady's tonier front-door company. (Nonetheless, Joanna did add a very comme il faut lobed fanlight above the door, with a tinted bull's-eye glass accent at the center.) Upstairs, the two elegant older parlors took on specific functions as library and guest bedroom. Travelers would no longer have to share sleeping quarters with the family, a reflection of the shift away from the communal living arrangements of traditional western Europe and toward the private accommodations Americans would come to insist upon.

The new and improved Manor Hall even had a kitchen that could keep pace with the Philipse family's demands for daily sustenance and frequent entertaining.

A well-equipped kitchen was essential for hosting a gathering such as the one Joanna was planning, which would be considered lavish by some standards but was typical for a well-to-do household that merged its personal and political interests during the festive winter social season. A hostess of Joanna's ilk would not need any excuse to throw such a party. There need be no special occasion for breaking out the fanciest plates, the finest silver. And for her, orchestrating the event came as no hardship. It was, essentially,

Joanna's job, a job at which she excelled and in which she took pleasure.

Preparations spread to every worktable. Heavy ceramic bowls cluttered the sideboard. Cutting boards, spoons, sifters, and graters passed from hand to hand. Even with all that space, the servant girl who emerged from the cobble-floored dairy balancing a pitcher of rich new milk and a plate of butter had to squeeze past the elderly, experienced cook who knelt practically inside the chimney to calibrate the hearth fires.

On an occasion like this, more hands joined the kitchen crew early in the day. Joanna's daughters were recruited to help—Susannah, at twenty-two soon to be married; Mary, a flirt of fifteen, for whom this party represented just one of the playing fields on which she would captivate the men of the colony; and Margaret, who at twelve was still more interested in licking the spoon and pinching raw dough than actually cooking.

Joanna drifted in and out to taste test dishes and give pots a ceremonial stir between checking on other preparations throughout the Manor Hall. She used to spend more time in the kitchen, but had graduated to observer status. She trusted her servants and slaves to execute her orders. And the lady of the manor was glad not to spend every day in the discomfort of the fireroom; displaying her cooking craft only on a special occasion such as this was quite enough for her. The mammoth fireplace poured out heat—the kitchen's temperature in cold weather hovered at around 100 degrees, shooting up in summer to 120—but also smoke, which stayed trapped in the room because the cook insisted on keeping those beautiful big windows closed. She needed to minimize drafts, which could render recipes less precise.

By the time Joanna emerged from the kitchen that morning, the sweets arrayed around the room wafted deliciousness. But her own hair and skin felt as if she had bathed in soot. Joanna would spend most of the afternoon in her bedroom dousing her body with potions, some she had prepared herself and others ordered straight from Paris. With all the rose water she threw around, she ended up smelling something like a marzipan hedgehog herself.

*I*t was party time. Joanna's confectionary buffet was a true spectacle. There were custards topped with blossoms. Sugared fruits mounded into pyramids and animal shapes. Cherries preserved in jelly. Dishes of preserved pineapple shipped all the way from the West Indies. Orange chips. Pistachio cream and raspberry cream. Her marzipan hedgehogs. A pastoral scene with sugar sheep tended by a sugar shepherdess. A pyramid of footed trays arranged with narrow glasses brimming with punch. And the whole business crowned with a pedestaled sweetmeat salver, a gleaming, glowing tower of glass ornaments. A triumph!

Of course, Joanna had not forgotten the savory to balance the sweet. Her epergne, an elaborate glass and silver serving dish with octopus arms, held delicacies such as preserved mushrooms, salty capers, walnut and tomato catsups, and cucumber pickle.

Joanna's spread was worthy of its place at a high-society party in 1745, guaranteed to incite the crowd's imagination as well as its appetite. Her guests would have sated themselves with their main meal in midafternoon, making this repast essentially their dessert course.

Most of the revelers arrived at sundown by horseback or sleigh. Sarah Kemble Knight, who rode to the city from Boston in the winter of 1704, marveled at the fifty- or sixty-odd sleighs she saw hurling themselves over the snows of Manhattan in a single afternoon: "some are so furious that they'll turn out of the path for none except a Loaden Cart." But it was not only thrills that made eighteenth-century travelers sleigh-crazy, or the cushy sensation of riding out in the frosty air, snuggled down in bearskin robes and swan-feathered muffs. There was literally no other way to get someplace beyond walking distance in the depths of winter. Some guests had traveled from lower Manhattan and planned to stay over, bunking together throughout the house. After all, the typical "assembly" of the time started at six and went on until nearly dawn. But the rest would leave their horses temporarily stabled while they consumed Madeira, raised dozens of toasts, and performed a stately minuet or a funky "English country dance" to the accompaniment of a hired violin and a pair of flutes. These cosmopolitan guests probably would have laughed at Puritan Pastor Increase Mather's pinched denunciation of the "gynecandrical dancing or that which is commonly called mixt or promiscuous dancing, of men and women." Joanna Philipse and her guests were New Yorkers, not Puritans, and they preferred their dances "mixt."

Throughout the evening, the door to the room with the banquet remained shut, in keeping with tradition, to make for a grand flourish when the precise moment for the display arrived. The hostess determined that moment, but it was expected to come when the festivities had reached a fever pitch—usually at about 11:00 P.M. Then the guests would rush in to feast.

Frederick and Joanna typically invited a few dozen political peers of Frederick's to a party of this size, along with some of Adolph's fellow merchants and as many of the local manor lords as they could summon. The wives, of course, flounced in wearing their finest regalia. Adolph himself would eventually appear, perhaps with a few of his bachelor cronies. Like Frederick and Joanna, many of the married couples were in their forties, and arrived with their own nearly full-grown children, adding a youthful luster to the gathering. Eventually the house would be almost too crowded, its rooms stuffed with fabric and flesh, redolent of smoke and sweat. Just as any person arriving would expect.

From the parlor's southern windows, guests could look out over the splashing falls of the Nepperhan and see the outlines of the hoary gristmill down over the shoulder of the hill in front of the house. To the south, across the bluffs that rose from the other bank of the Nepperhan, the Hudson River stretched toward New York, framed by its wall of umbrous stone, the Palisades. At sunset, as guests arrived, gold streamed over the top of the stone and across the water, casting its gleam over the assembly.

But the eastern exposure took precedence over any magnificent western sunset. From the eastern pair of windows, guests could view the turnpike that made its way past the Manor Hall from the island of Manhattan. When Margaret first erected this house on the Nepperhan, the road had been nothing more than a native trail. Then in 1703, when Frederick II was a child recently orphaned, the New York Assembly passed a statute for the establishment of a "Publick and Common General Highway to extend from King's Bridge in the County of Westchester," the bridge that belonged to the Philipses of Philipsburg Manor, "through the County of Albany, of the breadth of four rods, English measurement, at least." Known first as the Queen's Road, the highway came to be called the Albany Post Road. It was the cornerstone of north-south transit through the Hudson Valley.

By the early 1720s, the highway had been completed from Spuyten Duyvil north as far as Yonkers, where it eventually would come to be known simply as Broadway. Frederick found himself appointed to the lofty position of Westchester Commissioner of Roads in 1723, and he embraced the responsibility of executing his sovereign's mission. He expanded the road farther north, but also enlarged and graded the existing muddy, rutted highway, which ran less than a quarter mile east of the Manor Hall and which could be admired from Commissioner Philipse's house with few obstructions.

When friends came to visit, their experience in Joanna's parlor was as political as it was domestic. Her guests would recline in their fashionable spindle chairs around a little tea table (imported from the Continent) with Joanna's new matched creamware service, also imported (a switch from the old-fashioned collection of pieces acquired helter-skelter, the best of them honored heirlooms). Joanna's friends would fill their cups with the best cocoa from the tropics, yet another denominator of the family's status, as they peered out her glazed windows to see whose coach was passing through the trees screening the Post Road. Living on such an important thoroughfare itself made a political statement.

And a tea party paled in comparison with a grand assembly as an opportunity to show off both the house's address and the family's surfeit of table settings and cutlery. To set a table so amply—with enough spoons, knives, and forks for the entire company—rendered a visceral message about the Philipse fortune. Collecting household silver was the equivalent of keeping cash under a mattress in a time when no banks existed. The difference, though, was that an individual with cash under the mattress wants to hide and safeguard that money, and the approach of Joanna and her colonial peers was to trot it out, to flaunt the dimensions of their fortune. The silver on Joanna's buffet table represented both savings account and swag.

On an evening such as this one, the world came to the Philipse family. The Manor Hall glowed in the dark night, no small feat. In the fashion of the moment, Joanna and Frederick illuminated their home with sconces surrounding the fireplaces and paired with mirrors to throw light on the glamorous goings-on. One historian has noted the "old rule that called for at least one candle for every guest" during social occasions. Until now, wax tapers had proved exorbitant for all but royalty and the very top crust of aristocracy. Even the wealthiest New Yorkers made do with low light, cooking by a single candle, mending by a single candle. Looking glasses, however, could be bought cheap because new technology had created a more affordable type of mirrored surface. The looking glass magnified and reproduced the candlelight. The flames and their reflections danced like particles of the sun that had been cast about the room.

Rich finishes added to the impression made by the place. Sarah Kemble Knight, the eagle-eyed diarist of eighteenth-century New York, commented on the woodwork in the more well-to-do households, much of it "kept very white scour'd," scrubbed to bring out the natural grain, with furniture waxed and polished to show off the wood's luster. Plank floors were strewn with sand after scrubbing, then swept into swirls and other decorative patterns. The chemistry of paint had improved since the Dutch settled the area, expanding the palette favored for interiors. Whitewash over plaster walls was now joined by a Prussian blue lightened by white, a pigment so common one paint expert has called it "ubiquitous blue" (its popularity enhanced by the belief that the color repelled mosquitoes and flies). Microscopic analysis of woodwork in New York has revealed jaunty color duos: orange and teal, or lemon and brown, or a rich salmon that brightened up the traditional blue-gray. Cupboard interiors would get a coat of vermilion to accentuate the colors of the porcelain displayed within. (For baseboards the ever-practical colonists favored scuff-obscuring black.) Later would come yellow ocher and a bold grass-green (exacting Thomas Jefferson would obtain the perfect formula from a celebrated portrait painter for the hallway floors at Monticello) and one overarching favorite of the colonial era, Venetian red, a dark cherry shade with a hint of carnivale, a dramatic backdrop for the swaying fairy-colored layers of petticoats and the alabaster cleavage that swelled out of women's flat-front satin bodices.

In the candlelight it might have appeared as though black flies had lit upon some of the most beautiful faces, both male and female, in Joanna's parlor. But closer inspection revealed the shapes to be tiny hearts or crescent moons cut of black silk. These were temporary beauty marks known as "patches," whimsical gummed shapes to adorn the fashionable complexion. Patches also happened to be useful in hiding the pockmarks that afflicted so many in this era of smallpox contagion. Satirists lampooned the practice with comic renderings of fashion victims, faces obscured by inky blotches. In the mid-1600s England's chief moral cop Oliver Cromwell had tried to outlaw the decadent practice, but the fad persisted for well over a century, and the patches took on a variety of meanings. Commentators termed a patch at the corner of the eye an "impassioned patch," while one applied to the middle of the cheek suggested the personality of a "gallant." Patches even became political billboards: A Tory wore hers on the left side of her face and a Whig on the right. Both men and women carried patch boxes made of tortoise or china, brass or silver, ever ready to repair or replace a silk beauty mark as needed.

Joanna's guests shared her commitment to fashion, and her commitment—almost a sense of moral responsibility—to adding their feminine wiles to the amusement of the gathering. That meant conversing with wit and delicacy but also, first and foremost, adding a sheen of glamour to the assembly. So the women of upper-crust New York would go all out. Their costumes were the latest incarnation of a centuries-old European tradition of exaggeratedly sexualized women's clothing, which emphasized that all-important region between the waist and knee. With its female guests consisting mainly of hips and a grace note of bosom, the room might as well have been filled with cloth-draped fertility figures.

The women wore their dresses in shades of apple green, saffron yellow, powder blue, and rose, sometimes sprigged with silver and gold handworked blossoms, tucked or beribboned. By now, many women had adopted *le sacque,* the French gown then at the pinnacle of chic. With pleats that began at the nape of the neck and flowed down the back like a superhero's cape, the sacque's waterfall of fabric further reemphasized the thunderous dimensions of the widened skirt. Paired with one of the "pumpkin hoods" of the day, the look was not especially delicate, but eighteenth-century women wore the sacque with a passionate loyalty. It was universal for the women of society to accompany their gowns with long, sleek gloves in pastel shades—in summer they cut off the fingertips—and many would have perfumed the satin or kid with the scent of hyacinth or with a home brew of musk crushed with civet, rose water, and cinnamon.

On such an important evening Joanna would style herself a concoction of crisp, thin silk and lace, ribbon and silk and well-creamed flesh. Her lace-edged shift would peek above her tight-laced corset. The tops of her satin gloves stayed in place with the help of bands of woven horsehair, held snug beneath the flounces and tiers of lace that cascaded from her sleeves. If humanly possible, believed Joanna's peers, a woman must keep shrouded that most ungainly of body parts: the elbow. The lady's dress was stitched of quilted taffeta— gridelin is the color she ordered from her Parisian dressmaker, the perfect shade of platinum-gray-violet—and embroidered all over with vines and flowers in yard upon yard of silver thread. Its skirts hung just high enough to reveal the minxish, needle-nosed points of her emerald brocaded slippers. In the lobes of her ears she wore garnets that shone bright as rats' eyes. The stones were modest in size, yet picked up the flicker of the fire and the glow of all those candles. A fillet of ecru ribbon tied up her perfectly tousled hair.

Joanna Philipse's dress represented the latest Parisian novelty, the *panniere a coude,* usually simply called panniers. Her gown was supported by a double metal frame that extended her skirts so widely, by nearly a yard on either side, that she could rest her arms atop the silk shelves that jutted from her hips. Even with all the doorways recently widened during the house's renovation, Joanna and her female guests had to consciously strategize before moving from one room to the next. Some with wider panniers than hers had to make their grand entrance sideways.

The hostess was a vision, even at a well-seasoned forty-five. Women at middle age in colonial New York saw no reason to deprive themselves of the fashions that might to modern eyes more felicitously complement the figures of their daughters. It was a question of attitude. Joanna's personality comes through in an image rendered by society artist John Wollaston, whose visit to the city from 1749 to 1751 yielded a total of fifty commissioned studio portraits. Commissioning family portraits conferred status, and like-nesses by English painters hung in many elegant New York parlors. The painting shows Joanna wearing a shawl-like fichu studded with a fat satin bow, her hair covered by a simple white frill of a cap. Her countenance—the penetrating eyes, arched eyebrows, slightly pursed, hint-of-a-smile lips—seems the perfect wry affect for a high-style political hostess.

For Joanna and her contemporaries, fashion came directly from Paris, literally as well as figuratively. Each season, the fashion babies arrived by sea and made the rounds of English society on the Continent and in the colonies. Clad by client-hungry dressmakers, these hand-high dolls modeled the current couture from head to toe; affluent women then ordered new dresses in their choice of design and fabric. The London milliners, suppliers of dry goods rather than hats at that time, also sent fashion babies to America with apparel made from their textiles, in the event that consumers preferred to purchase bolts of fabric and have local seamstresses finish the work. After the fashion season, the figures were recycled as dolls for privileged little girls.

Apricot face creams, lipsticks, and all varieties of rouge also could be ordered from the Continent, and some of Joanna's peers considered them indispensable. Since the time of Elizabeth I, who slathered her aging face with concoctions she herself invented (rendered "puppy dog fat" was a key ingredient, along with snail shells, butter, and lemon), ladies had used complexion creams as a tonic for their weathered skin. It is easy to imagine the damage they incurred with the most popular potions, including white face powder made principally from lead, since the fumes emitted during its preparation were known to blind the workmen who prepared it. Heavy use of perfume was de rigueur—anything but resort to soap and water. (Some women made an exception to cleanse their faces with "wash balls" imported from the Netherlands.) The mid-eighteenth century was not a natural time. Women sometimes supplemented their eyebrows with shapelier ones cut from mouse skin, and were known to stuff their cheeks with small cork balls called "plumpers" to present a healthy-looking face to the world.

*T*he male guests who cavorted at the Philipse house, most of them members of the council, merchants and gentlemen, displayed their own brand of glamour, as men had since European explorers first stepped on American soil in their fully doubletted regalia. Like their female companions, New York men focused rapt attention on their garments, often made of fine taffeta and other silks, gold-encrusted fabric, and voluptuous quantities of lace. For a time men's coats had stiffened into flaring skirts, some with pleats, until the male silhouette evoked the female pannier. Gentlemen left their waistcoats unbuttoned to show off their ruffled cambric shirts, a practice many women viewed as rakishly sexy. (According to the *Tatler* of 1710, "A Sincere heart has not made half so many conquests as an open waistcoat.") Lace gushing from wide-cuffed coat sleeves showed that a person was a gentleman, not a worker. Affluence, as always, was the greatest aphrodisiac of the time.

All men of Joanna's time and class wore elaborate wigs as a central part of the wardrobe; those who shunned perukes were hopelessly eccentric, no matter their class. Benjamin Franklin, for example, became known for his idiosyncratic fashion sense. A 1777 engraving depicted his distinctive wire spectacles, marten-fur cap, and shoulder-length, unpowdered locks that set him apart from most men of his age as surely as his genius. Even children, slaves, and convicts regularly wore false hair. In part, men wore wigs over shaved pates as a hygienic replacement for long, natural hair, which was nearly impossible to keep clean—particularly when the average citizen almost never shampooed. Those who could not afford a wig woven from human hair sometimes wore horsehair or goat hair, and the most impoverished were known to sport wigs constructed of thin wires. Throughout the British colonies, as in England and the rest of Europe, perukes and the bushier periwigs stayed in style for more than a century, except for the period around the Great Plague of London, in 1665, when people feared to wear anything that might have been constructed from the hair of victims.

Most men at Joanna's party had, like her, spent a good part of the afternoon preparing themselves for the social engagement. That meant climbing out of the flowing chintz dressing gowns called banyans and removing the embroidered caps that covered their shaved pates at home, to get their most fashionable headgear in gear. City dwellers of the period describe Saturday afternoon streets crawling with delivery boys bearing wigs to gentlemen from the shops of barbers who had unstintingly heated, rolled, brushed, and styled the crucial accessory. After donning the wig, its wearer cupped a paper funnel to his face so that he could breathe while a servant lavished on his head up to two pounds of hair powder, consisting of fine-ground starch or wheat flour or sometimes plaster of paris (with more than a hint of orange flower scent, usually, and perhaps a tint of violet, pink or blue). Eventually, the best homes had a separate room for this ritual, called a powder closet (undoubtedly the forerunner of the petite but functional powder room).

The most dashing wig style at the moment of Joanna's party incorporated a "crapeaud," more casually called a "bag," a dark silk pocket at the nape of the neck into which a gentleman channeled his fake hair and tied it with a fancy bow. Styles abounded, with names reflecting the creative energy that fueled the wig fetish. There were "pigeon wings," which flaunted rolled hair-puffs on either side of the face. The "full-bottom" made its statement by flaring over the shoulders in front and back. The "club" featured ends curled under pageboy style toward the wearer's neck. Barbers trumpeted their expertise in advertisements touting "the comet," "the cauliflower," "the staircase," and "the Adonis." But the "ramilies" was the longtime favorite, recognizable from eighteenth-century paintings, with its pouf of hair above the head, braid in back, and black ribbons embellishing both. Sometimes a man wore another black ribbon around his throat to secure the braid in back, a flourish called a "solitaire" that the fashionable viewed as a particularly natty grace note. The large comb that protruded from the gentleman's coat pocket was used, publicly if necessary, to put his complicated wig in order.

Men of Frederick's class also paid keen attention to the cut of their trousers, which had grown steadily snugger since the days of New Amsterdam. Then, Dutch settlers wore their sheepskin pantaloons comfortably baggy. Though the codpiece had vanished from the fashion scene, the New York stud now sported breeches and stockings that were glove tight, and assumed a continual balletic fourth position to best display the contours of his flexed calves and thighs, which were linked to a man's sex appeal as inflated hips were to a woman's. One colonial raconteur cites a gentleman ordering from his tailor: "If I can get into 'em, I won't pay for 'em."

The average New York swell wore a corset, yellow kid gloves, worsted hose, silk pumps, marcasite shoe buckles, and chicken-skin fan. His peers matched him garment for garment, accessory for accessory. Dress was a major badge of rank. So gentlemen indulged in whatever ornaments they could afford. (Men's wardrobes weren't all fey: For a martial look colonists wore moose-skin coats; another popular garment was the tough-sounding "thunder and lightning coat," and stiff, tall, black top boots had a macho appeal.)

Men and women expected an elaborate degree of choice in their haberdashery. Shoppers browsed for textiles in catalogs, which listed hundreds of different weaves, weights, and blends. These people were particular about fabric. Consumers of later centuries recognized material simply as cotton, linen, or rayon, but Joanna would have known the distinctions between dozens of varieties of lutestring (fine silk), armozine (strong corded silk), baize (coarse wool), bombazine (silk and cotton), callimanco (glossy wool), cypress (cobweb-thin silk for mourning clothes), sergedesoy (coarse silk), dimity (ribbed cotton), dornex (heavy coarse linen for servants' jackets), erminetta, fearnothing, ferrandine, gingerline, humhum. The list goes on.

There was a tactile poetry to dressing. It extended to the patterns woven into the fabrics, some of which sounded like scraps of the new romantic novels, such as Daniel Defoe's 1722 *Moll Flanders,* Samuel Richardson's 1740 sensation *Pamela,* or Henry Fielding's *Tom Jones,* that ripped through literary society in 1749. ("Today it is our pleasure to be drunk," averred Fielding, skewering his midcentury peers' polite but total immersion in matters of the flesh.) One typically romantic fabric name, in this case a variety of calico, was "harlequin moth."

Colors, too, seemed imbued with emotional flavors. Lutestring could be found not only in pink, plum, and cinnamon but also in the perhaps questionable "flystale." People identified clothing color by other things they knew well, so "milly" was slang denoting meal-colored, and "mouse-colored" muffs must have had a tinge of common rodent. Popular colors were decidedly less muted than they had been in the previous century. Colonists showed themselves in clothes of scarlet, fizzy sherbet colors, "waistcoats of grain" (another word for red), and crimson damask petticoats.

Accessories received minute attention regardless of the wearer's gender. Buttons were crafted of coins, paste, or colored enamel. Others were fashioned from mohair, steel, gilt, stones such as lapis lazuli or agate, or even a chunky yellow horse's tooth set in thick brass. Gentlemen cast their own pewter buttons in molds at home. Even buttonholes made a statement, often featuring elaborate ornamentation of metallic thread. Buckles for shoes and men's knees had paste jewels set in gold and silver. Ribbons, termed "love ribbons" in advertisements, made popular gifts for St. Valentine's Day, and adorned dresses, chemises, jackets, and the hair of both men and women.

By midnight, the parlor of the Philipse home overflowed with men in their loin-hugging breeches pressed up against women's rustling silk hoops, which they liked to tilt up just enough to expose what the English poet Robert Herrick suggestively termed the "tempestuous petticoat." It was always a gamble whether the capacious hoop would stay in place or crumple upward like a collapsing Japanese lantern to reveal the wearer's naked thighs. Englishman Robert Campbell published this praise of hoops in 1747: "We see they are Friends to Men, for they have let us into all the Secrets of the Ladies Legs, which we might have been ignorant of to Eternity without their Help; they discover to us indeed a Sample of what we wish to purchase, yet serve as a Fence to keep us at an awful distance." Predictably, one popular rustic amusement was for men to push bare-legged, loose-skirted young ladies high in the air on swings. (To go drawer-less beneath your street clothes was a habit shared by men, say the principal authorities on early American dress, explained by the fact that comfort required nothing else, since breeches were customarily sewn of "such sleek materials as silk and satin." They elaborate: "That no needlessly encumbering materials were worn by the adult [male] is made abundantly clear in eighteenth century diaries for entry after entry bears witness to the instant availability of the sexual apparatus.")

Female guests had been invited upstairs to visit the newly painted chamber that was designated tonight for their refreshment. (The room served multiple purposes, including sleeping.) In one corner, a tall cage of twisted saplings held an apricot-colored, fluffy-feathered Moluccan cockatoo, a souvenir brought back from Indonesia on a merchant ship. A second cage held an American mockingbird that the family had captured just outside the house. The bird had not yet let loose its jazzy arias in captivity, but the family nonetheless enjoyed the bird as a souvenir of the tamed wilderness, like the squirrels colonial boys liked to keep on leashes. A small mahogany table in another corner of the chamber held a ribbon loom, which offered a graceful activity for the girls while they entertained friends and suitors. (A Philipse could of course purchase any quantity of "love ribbons" she wished and did not need to actually produce anything at all.) There were small upholstered chairs on which to perch in front of the fire.

Servants also brought in chamber pots for each woman's relief, and waited at silent attention by the side of the room to take them away. They did not suffer any discomfiture, these superfine ladies of New York, as they squatted down one after the next to do their business in the lit-up room; after all they had the cover of their puffs of petticoats. And it was so much more comfortable to void here by the cozy fire, even in company, rather than risk a chill in the outhouse, called the necessary, that was shadowed by the bushes near the drinking well. Anyway, using the necessary tonight would mean ruining a pair of satin dancing slippers in the snow, and it could take six months before a new pair waltzed its way across the water to New York.

Downstairs, the men brandished the pipes that had been offered, and engaged in the usual political discussions while they blew richly acrid lungfuls of smoke out Joanna's new sash windows that were raised high to let in the midnight air. As for the ladies, only the old Dutch women who didn't care how they looked smoked nowadays.

When all the guests reassembled in the parlor, a liveried slave emerged from the kitchen door with a large pitcher in his hand. With everyone watching, he lifted his arm slowly and dramatically, pouring a cascading stream of warm cream into a pot of sack and so creating foam and curds that would be flavored with lemon juice and sweetened with powdered sugar. Syllabub was a brew with festive clout that equaled its alcohol content. Its popularity extended beyond the wealthy—the farmer down the way might even milk his cow into a bowl of hard cider and call that homely broth a syllabub. No matter how it was prepared, however, one element remained a requirement: the crown of bubbled whip atop the liquid in each drinker's cup. Inevitably that foam on the syllabub left guests pleasantly sated and snockered at the same time. A party might grow louder, then, even raucous—turning into the kind of gathering the English termed a "rout"—as the veneer of a refined assembly crumbled away like the last bits of marzipan on a plate.

13

A Hard Winter and Hell

*C*onsidering the effort it took to get anywhere at a rate of five miles per hour, particularly without graded highways, a wealthy colonist in 1740 could spend a lot of time closed up in an elegant, teeth-clattering box. A coach maker had only recently opened in New York, meaning that nearly all of the modish carriages driven by the best families had to be shipped from Europe. (The sight of a gleaming coach being winched ever so carefully out of a ship's hold in New York Harbor must have drawn dockside crowds.) Even after half a dozen manufacturers set up shop in Manhattan at midcentury, a carriage remained an exorbitantly priced status item that few could afford. Coach production integrated the precision and creativity of tanner, blacksmith, machinist, carpenter, woodworker, wheelwright, and painter. Customers paid dearly for all that expertise. Painting techniques alone advanced so much as the trade evolved that it was not unusual, ultimately, for a high-end gig to glide (or, more to the point, bump) through the streets of the city bearing a Maserati-worthy finish of fourteen coats of lacquer green.

But no matter how splendid the finished product might appear, many well-to-do New Yorkers considered their carriages incomplete without a full complement of enslaved black coachmen and footmen—in white periwigs, liveried to the nines, and ever ready to serve. The uniform of livery spoke an instantly understandable language: The woolen coat, waistcoat, and breeches always included two contrasting colors, preferably the colors of the family coat of arms (ancient or concocted more recently), elaborately trimmed with woven velvet borders called "livery lace" and with cast buttons that also featured family crest motifs. (Washington and Jefferson, both gentlemen, attired their slaves in family colors and livery lace.) Attiring coach attendants in formal wear served to trumpet the owner's affluence in a way that anonymous buckskin breeches and a plain flannel jacket could not. For much the same reason, it became fashionable for the aspiring queen of fashion— someone like Joanna—to accessorize her look with a small black boy in uniform serving her at home and out and about. Beyond the symbolism of the livery, his dark complexion was thought to show her ivory skin to maximum effect.

To pass the time in a coach, passengers might imbibe, dine, slumber, make conversation, or even indulge in some friendly silk-sheathed frottage behind the taffeta curtains. The tradition of meeting cute in an overland coach was already fixed later in the century when Laurence Sterne published his international smash hit, *A Sentimental Journey Through France and Italy,* in which the deus ex machina for the entire tale is the dilapidated secondhand carriage Sterne's narrator purchases for his seduction-focused road trip. But not all colonists traveled as passive (or preoccupied) commuters, their progress dictated by coachmen. Joanna Philipse, for one, earned a reputation as an able horsewoman, and her performance on her four-in-hand—a carriage pulled by four horses—as she steered her jet black stallions across the bluffs of Philipsburg already was a matter of legend.

$\mathcal{R}$ecreation would have ranked high on Joanna's list of priorities. She was different from hardworking Margaret, who would not have bothered with frivolous entertainments, and earnest Catherine, who was not the embodiment of joie de vivre. Joanna's own husband had been instrumental in serving the long-term recreational needs of the city when he helped to secure the quarter acre that would become Manhattan's first public park. In 1732 the Common Council had voted to lease a small tract at the foot of Broadway just north of Fort George before all the land in the South Ward was completely given over to development. Then came the big guns. Frederick and four other appointed committee members were assigned to do the legwork on the project. The men's appropriately arduous assignment consisted of strolling a few blocks from their cushy homes to "lay out the Ground at the lower end of Broadway near the Fort for A Bowling Green" (the sport of lawn bowling being a perennial favorite among both the Dutch and the English), which would require that they "Ascertain the Demensions thereof with the breadth of the Streets on all sides." And so they did. (And likely shared a bowl of rum punch afterward as compensation for their exertions.) In 1733, once the park had been satisfactorily sodded with grass and enclosed with an iron rail, Frederick and two of his fellow park developers assumed a ten-year maintenance lease on it at the unbeatable rate of one peppercorn per annum.

At the time of the Bowling Green's development, Joanna and her children were still spending most of their time in Manhattan, where they could enjoy any number of diversions. A map drafted in 1735 by an amateur cartographer who signed her work "Mrs. Buchnerd" offers an indelible record of the entertainments that defined the city in Joanna's lifetime. The first known plan of Manhattan to be executed by a woman, it detailed the popular haunts that mattered to citizens such as herself.

The map detailed the drinking establishments known as "Meed houses"; the pleasure garden called Vauxhall—in the future Tribeca, occupying Greenwich Street between Warren and Chambers—where guests could dance, dine, and stroll though sylvan landscapes; and three similar "resorts," including a place named the "Spring Garden" and an ornamental retreat she labeled "the winyerd." She highlighted the Horse and Cart on William Street, the most popular tavern on a street of taverns. A fashionable "Coffy House" was recorded near Dock Street. On a more practical note, the four major markets are identified (the Fly Market, the Meal Market, Coenties, and Old Slip). Along both the west and east shorelines, notations read "Fishing Place." On either side of "Bowre Layne"—the boulevard that pushed northeast through the wilderness, past the up-island country estates of the elite (Stuyvesant, De Lancey)—Mrs. Buchnerd rendered the symmetrical pools that were famous as sites of pastoral idylls, Buttermilk Pond and Sweetmilk Pond.

With all its routs and frolics, its "Coffy Houses" and "resorts," New York had certainly become *très amusant* at midcentury. Young people went to all-day turtle feasts that took place at "houses" along the East River. Reverend Andrew Burnaby from England expressed his delight at these flirtation marathons, where "thirty or forty gentlemen and ladies meet to dine together, drink tea in the afternoon, fish, and amuse themselves till evening, and then return home in Italian chaises—a gentleman and lady in each chaise." Couples en route had to cross a tempting diversion, the Kissing Bridge, which arced over a picturesque millstream just about where the chamfered crown of the Citicorp Tower later would stand in midtown Manhattan. There, the vicar noted slyly, it was "part of the etiquette to salute the lady who has put herself under your protection."

Unprecedented novelties were arriving on American shores, events glowingly described in the publicity notices that appeared in the *New-York Gazette* alongside foreign news items and lists of ships that had entered or cleared the port. Recreation was morphing into an art form, at least for the elite. Music societies sprang up. People attended pleasure gardens to watch pyrotechnical displays, already known as fireworks. The only colors whose formulas had been devised at that point were yellow and orange, but that did not stop the production of huge displays known as set pieces, which sometimes depicted royalty and other popular figures to the accompaniment of orchestral music.

New York opened its first theatrical playhouse on Nassau Street in 1732 in a warehouse belonging to Acting Governor Rip van Dam. (The first play's male lead was acted, in the words of the single theater reviewer, by the "ingenious . . . barber and peruque maker" of Mr. van Dam himself, the aptly named Mr. Heady, with barber rather than waiter a common day job for aspiring thespians.) Shakespearean tragedy would arrive in 1750, when a visiting troupe from Philadelphia staged *Richard III* in the same carnelian jewel box of a dramatic space (red being the ubiquitous paint shade for theatrical interiors then) seating only a few hundred spectators, with a green velvet curtain, wax footlights, and an orchestra of German flute, horn, and drum. Above it all swung a chandelier made of a barrel hoop stuck with a half-dozen candles.

Clever mechanical devices usually took precedence over live action. Mr. Pacheco's Ware-House on Petticoat Lane, for example, touted a musical machine that performed "several strange and diverting Motions to the Admiration of the Spectators." During the show, doors flew open, men yanked bell ropes, chimes sounded, and a barber scene ended with all the ringers clean shaven, an event presented "entirely by Clock-Work in imitation of St. Brides Bells of London." Patrons came to a down-in-the-heels neighborhood nearby to study wax figures of the royal family of England. Entrepreneurs opened their homes to display natural curiosities such as porcupines, alligators, and electrical fish, while at the King's Arms Tavern at the Bowling Green, a "very beautiful" tiger found an avid audience.

A myriad of sporting events kept the more affluent men of New York entertained. Yachts and pleasure boats docked at every country seat with water frontage. Gentlemen invited their colleagues to shooting parties in their private deer parks. Bullbaiting, imported from England, pitted specially bred bulldogs against maddened bulls; the sport took place in Manhattan by the Freshwater Pond at a well-known rise in the land called Bayard's Hill.

An enthusiasm for horse racing showed in wagers placed on the speed with which a rider could take the fourteen miles from the city gates to King's Bridge and back (one rider made it in one hour, forty-six minutes). Hempstead Plains on Long Island (later Belmont Park) still drew the greatest crowds, despite the trouble it took to get there; at one race in 1750 an astounding one thousand horses competed, according to a newspaper account. The wealthiest horse lovers bred their own: Both the Morris and De Lancey families kept renowned stables in Westchester County, where they maintained their own manors neighboring Philipsburg.

Some accomplished men spent so much time patronizing coffeehouses by day and eating out night after night, it is hard to imagine how they made the money to afford it all. Frederick's uncle Adolph, for example, dined regularly with a klatch called the Hungarian Club at Todd's Tavern, which came under the observation of diarist Dr. Alexander Hamilton as he passed through the city in 1744. Hamilton detailed the bumpers of strong drink and the backgammon matches that preoccupied the hard-driving merchants, physicians, ministers, and judges assembled for supper, along with the delicacies they downed: "veal, beef stakes, green pease, and rasp berries for a desert."

More cleared roads enabled elite families to flee the hot city for the rustic pleasures of their country retreats. In summer, Joanna's destination, Philipsburg, was distinguished by both its pastoral land-scape and its hardworking farmers. The same Dr. Hamilton also stopped at a farmstead in Philipsburg; leaving his sloop, he ventured east to the Post Road and found a tenant farmer fashioning a sleigh. He provisioned Hamilton with "three fat fowls for ninepence and a great bucket full of milk into the bargain." Real estate ads open another window into Philipsburg Manor life at midcentury, with properties that sound opulent but were small potatoes compared with all that the Philipses possessed. One near King's Bridge boasted a five-bedroom farmhouse, "20 Acres in Orchard" and "200 Acres of Mowing, Plow, and Pasture Lands, together with 80 Acres of woodland, well timbered."

At the Manor Hall, the most prestigious address in all the manor, Joanna snipped blooms in the formal flower beds Frederick had ordered to be prepared, alongside a line of horse chestnuts that sent their prickly seedpods plummeting to the grass every October. Joanna—her sensibilities honed by the fashions of the city, by its offerings and indulgences—would want a beautiful garden, no less than Frederick, who had seen firsthand the sculpted landscapes of the great English estates. The couple now vied to produce an equally fantastic groomed paradise of their own where there had always been forests and wildflowers.

*B*ut time for play would not last long.

Frederick Philipse II had traveled the Albany Post Road south from Philipse Manor Hall to the town of Eastchester early on October 29, 1733. Eastchester, a well-populated farming community of Protestant dissidents, stood just west of the boundaries of Philipsburg, within Westchester County, where Frederick was the largest landowner. It was election day, and Frederick had a proprietary interest in the outcome.

On this much-anticipated morning, men already had assembled

in front of the small, square wooden meetinghouse known as the Church at Eastchester. The voters were ready to choose a Westchester representative to the New York General Assembly. Frederick had thrown his weight behind a little-known candidate named William Forster. His key qualification was his employment as a schoolteacher by the Society for the Propagation of the Gospel in Foreign Parts.

Forster's opponent was landowner Lewis Morris, lord of the Manor of Morrisania and the Philipses' next-door neighbor in the Bronx (part of Westchester County until the second half of the nineteenth century). His motives for entering the race were political rather than religious. He wanted to settle an increasingly rancorous score.

Two months before, the province's new governor had banished Morris from his post as chief justice of New York's Supreme Court, where he had served alongside Frederick and James De Lancey since 1715. Governor William Cosby, an inexperienced and demonstrably venal English import, had dismissed Morris after the justice refused to hear a case involving the governor himself. Cosby had hoped to sue the province's interim governor, contending that he deserved half his predecessor's salary.

The firing of Justice Morris angered many New Yorkers. Morris decided to retaliate by challenging Cosby's choice for the assembly seat, the schoolteacher Forster.

As election day dawned and Frederick rode onto the village green, he observed hundreds of men circling on horseback, preceded by trumpeters, violinists, and people carrying banners expressing support for Morris and critiquing the government.

After Morris's riders dispersed, Frederick and Justice De Lancey escorted Forster onto the field. Again hundreds of horsemen paraded around the square's perimeter, though this time they were Forster's supporters, augmented by dozens of Philipsburg farmers recruited by their manor lord. Under ordinary circumstances they would not be permitted to cast a vote, since the franchise was limited to freeholders who owned forty acres of land. But the rules had been amended for this election and tenant farmers who cared to take part need only their tenant farmer's lifetime lease plus forty shillings.

Finally, High Sheriff Nicholas Cooper, mounted on a steed with scarlet housings, holstercaps, and silver lacings, arrived to summon voters to the green. Frederick Philipse still assumed that victory would go, as it always did, to the conservatives, to the Church of England, to the government party. To him.

He was wrong.

The tally came in at 231 to 151, with Morris the victor. The result was all the more astounding considering the polling method. With literacy far from universal, paper balloting and the anonymity it provided was not yet an option. Instead Sheriff Cooper stood in the center of the green with a candidate on either side. After Morris and Forster each delivered a speech, the sheriff called for voters to "stand by your man"—literally to line up behind the man of their choice, so that the votes could be physically counted.

The 1733 assembly upset created a perceptible fissure in the durable wall of privilege upon which the colony and the Philipse mystique rested. It would take forty more years for the men and women of Philipsburg to find themselves without a lord and Lewis Morris's namesake grandson would go on to sign the Declaration of Independence. But even now, it was clear that the tectonic plates of class and politics in English America were beginning to shift.

The Morris coup emboldened the Popular Party and fueled its appeals for new electoral procedures, such as instituting the annual assembly elections that already took place in Boston and Philadelphia (in New York elections were occasional affairs, dependent upon the whim of the governor), and rejiggering representation to reflect population growth. The slate of popularly elected officials should be broadened to include mayors and sheriffs, said Morris and his associates Rip van Dam, attorney James Alexander, and Cadwallader Colden (he who had nothing but contempt for the legal know-how of Frederick Philipse). They even backed the shocking concept of a secret ballot. No more Stand by Your Man.

But first they needed a venue to popularize these ideas. At the time, New York had just one newspaper, the *New-York Gazette,* which automatically published whatever Governor Cosby wanted printed and ignored any opposition. To serve as an alternative voice, the Cosby critics founded a second newspaper, the *New-York Weekly Journal.* German immigrant John Peter Zenger launched the newspaper from his print shop on November 5, 1733, and the premier issue was devoted to the Eastchester election. Zenger, a Dutch Reformed congregant who had arrived with his widowed mother and sister in the early 1700s, was single-handedly responsible for keeping Dutch literature alive in the colony, publishing twenty titles—Bibles, Psalters, hymnbooks, almanacs, and textbooks— during his residence in America.

The independent-minded *Weekly Journal* had immediate success throughout the province. Colonists relished its clever skewering of Cosby. But the newspaper went beyond satire and championed the virtues of an unfettered press and free speech for individuals, radical ideas for the era. (Some of the *Weekly Journal*'s pithier bits found their way into Benjamin Franklin's and Samuel Adams's works supporting the cause of American liberty.) In pre-Revolutionary, even pre-pre-Revolutionary New York, Zenger may have been appreciated by the public but he was infuriating to the powers that be. The governor even ordered the public hangman to burn copies of the newspaper in front of City Hall. The hangman refused. Finally Zenger was arrested on charges of "seditious libel." Unable to raise bail, he languished in a jail cell for nine months before his trial began. (The anti-Cosby flame stayed lit, though, during Zenger's incarceration, due to the efforts of Zenger's wife, Anna Catherine, who served as publisher throughout and did not miss an issue.)

Frederick and fellow justice James De Lancey presided over Zenger's trial, which drew overflow crowds each day. The defense attorney for Zenger, a Scotsman named Andrew Hamilton, rode north from Philadelphia, where he was known not only for his legal skills but also as the designer of the Pennsylvania Assembly's new building, Independence Hall. "It is not the cause of a poor printer," he reminded the jury, "not of New York alone, which you are now trying. No! It may in its consequences affect every freeman that lives under a British government on the main of America. It is the best cause. It is the cause of liberty."

The royal judges, unshaken by Hamilton's eloquence or the excitement of the crowd, urged conviction. The jury voted for acquittal. Citizens roared with approval for the verdict and some disapproval for the two justices on the bench. Frederick found himself on the wrong side of his fellow New Yorkers.

A second incident to embroil the Philipses would later come to be known as the Great Negro Plot of 1741, but it was actually a multiracial affair that pitted New York's disgruntled underclass against the city's property-holding, pleasure-seeking elite. The eruption of hostilities and destruction shocked citizens who had managed to turn a blind eye to similar incidents in the past. And the turmoil hit especially close to home: One of the instigators was a Philipse slave. And once again Frederick had been called in to prosecute those involved.

This plot was preceded by a natural harbinger of disaster even more alarming than a comet. The five-month period between

November 1740 and March 1741 would long be remembered with a simple phrase: the Hard Winter.

A New York winter was rarely anything but frigid, dark, and overlong. But the Hard Winter of 1740–41 took its place in the annals of American weather as one of the region's most miserable. Not that the season's extremes were recorded with any scientific precision. The tools to mark and analyze temperature and precipitation were only just coming into existence. It was as recently as 1714 that Daniel Gabriel Fahrenheit had found a way to gauge temperature variations using a column of pressurized mercury in glass. In the nearly four decades since, his brainstorm had remained a novelty, and thermometers had not become part of the universe of ordinary Americans. Whatever comfort comes from organizing the perception of natural experience into quantifiable terms—controlling at least the idea of arctic domination by transforming it into a confined, discrete string of single-digit days—was not available to the people of Manhattan.

But it did not take scientific tools to grasp the toll of human deaths from exposure, livestock lost in snowstorms, or the month and a half that the New York Harbor sat frozen, locking out the ships laden with the supplies city residents so urgently needed. By mid-January the rivers, channels, and bays were closed, even salt estuaries that had never been known to freeze. Scotsman Francis Lewis (another future signer of the Declaration of Independence) rode a sleigh two hundred miles from Barnstable on Cape Cod to New York City along the Atlantic's usually crashing ocean shoreline.

During the Hard Winter, a blizzard that dumped snow as high as a man's waist on Christmas Day was followed by a series of northeasters that pounded the Atlantic seaboard. The snow continued until it stood shoulder-deep in the snowplow-less streets of Manhattan. White drifts scaled buildings, shrouding second-story windows and smothering the sounds of street traffic. Horses floundered in snowbanks. The endless snow, frozen in place then piled with more snow and crusted with sleet, paralyzed the city. People found themselves homebound, as movement around and in and out of town nearly ceased. Even staying home was no remedy for discomfort, though, because wood and coal were scarce and costly. "The cold is so excessive, that while I am writing in a warm room by a good fireside, the ink freezes in the pen," reported a correspondent for the *New-York Weekly Journal* on December 22, 1740. In outlying areas of upper Manhattan and Westchester, farmers came upon herds of deer lying dead, starved, their carcasses frozen into the snow. The wolves had picnics all winter long.

The poor of New York endured disproportionate hunger, as did enslaved men and women and their children. Residents had no way to get fresh provisions. Only those households with a stockpile of goods could be sure where their next meal would come from. The hardships grew worse when a long-simmering European war finally broke out—England against Spain—and fueled further anxiety about food and security. Then the price of wheat shot through the roof. Bakers notified their clientele that they would not be stoking their ovens until they could obtain cheaper flour. The grain shortage even affected the affluent.

Not, of course, the Philipse family, for whom a want of bread would be as unimaginable as a lack of oxygen. Wheat abounded in the granaries of Philipsburg. If Joanna felt a pinch of cold, she demanded hot tea and it would be brought to her directly. She could pile her overstuffed, linen-swathed feather bed with as many thick wool blankets as she desired when the windchill dropped below zero, and there were still more blankets stacked in the lumbering old family *kas*. She did not even have to rise from her bed to grab another coverlet; a slave would get up off the floor to tuck it in around her.

Nonetheless, as the cold began to retreat, difficult questions weighed on Joanna as they did on the rest of the city's devout citizens, privileged and needy alike. Notable natural events—the usual comets, droughts, wheat blast, beached whales—always provoked soul-searching. It would be nice to accept the exigencies of winter as God's generous gift of essential moisture for the preservation of men. But what if God was in fact angry? What if, as one reverend phrased it in a sermon on March 26, 1741, God actually intended that "the Air that he gave us to live and breath in, should become the instrument to execute his vengeance on us, for our ingratitude to his goodness, and our transgression of his law"? In conversations the faithful plumbed the meaning of the scriptural question, "Who can stand before his cold?" Did this unprecedented freeze constitute a punishment for past transgressions? Or did it foretell horrors to come?

*T*he winter had not yet passed when the first inkling came that something else was wrong in lower Manhattan. A little after noon on March 18, 1741, residents heard the chimes of the city's fire bell. Men and women hustled to the foot of the island to find Fort George ablaze. High winds whipped the flames out of control. The structures inside the compound—the royal governor's house, colony administrative offices, armory, barracks, chapel, and provincial secretary's office—had already caught fire or would imminently. Men dragged the city's two hand-pump fire engines all the way from City Hall and the bucket brigade went to work, but the flames quickly climbed to the rooftops. Residential lower Broadway and Stone Street looked to be next, so residents attempted to rush away their valuables, slogging their china, paintings, and even heavy rugs through the smoke and mud. If the wind had not abruptly shifted, the Philipse family would have lost the family home hard by Fort George.

In the end, the fort and all its inner structures were decimated, left in ashes, and all the provincial records destroyed. And no one could imagine how the fire started.

Then a rash of mysterious fires broke out, their causes undetermined. For three straight weeks, the fire bells sounded nearly every day, sometimes twice in one day. Warehouses and residences burned all over town. Property owners did not know when they woke in the morning if they would be bereft at nightfall. Next came a new twist: Word spread that in some cases, robberies had preceded the fires.

In April the Common Council announced the start of a full-scale investigation. It offered rewards to encourage witnesses to come forward with information—one hundred pounds for whites and twenty pounds for slaves, which though clearly inequitable represented a formidable bounty to black New Yorkers, especially since it included a promise of manumission.

The fires served to heighten the tensions of the Hard Winter. And then something else began to boil to the surface, an idea that would make sense of everything—the cursed winter, the string of probable arsons, the reports of black-on-white thefts. The idea had its spark, actually, on the afternoon of the very first blaze, when a cache of grenades in the fort exploded. For a terrible moment the crowd recognized the sound not as a freak accident but as the opening salvo in a war fomented by an army of firebug black insurgents.

The fear that rose in many white New Yorkers came in the wake of years of talk about slave rebellions in the other English colonies. There were a lot: The premier historian of slave rebellions in America put the total number of uprisings at 250, a low-ball estimate because it excludes the scores of ship revolts and cabals of less than ten. The Fort George grenades resonated as gunfire because people knew that guns had sounded many times before, as close to home as Boston and New Haven. New Yorkers in the first half of the eighteenth century had also heard much talk about Virginia. In one plot, runaway slaves sequestered themselves with guns and ammunition in Virginia's Blue Ridge Mountains until a vigilante army brought them back into bondage. Even attendance at funerals of fellow slaves afforded stealth opportunities for Virginia dissidents to conspire. White colonists' churchgoing became a strategic opportunity for rebels in 1710 when hundreds of blacks and Indians planned an armed Sunday morning surprise, and the reverent filed out from services to find their world rocked in the most brutal fashion.

Two years earlier in South Carolina, just outside Charleston, another rebellion had become so famous that the name of its location, the village of Stono, became synonymous with armed resistance. King Philip of Spain, the colonizer of Florida, had announced that he would free every English-owned slave who fled to St. Augustine—an ironic offer coming from the leader of one of the main propagators of the slave trade, but politically expedient as England and Spain prepared for war. In 1739 Philip upped the ante with the advertisement of a settlement to be founded by and for escaped blacks. Soon word spread among slaves throughout the south: If they could make their way to Florida, they would be welcomed by a village of freedmen (called Fort Moosa) where all residents would be armed against their former masters.

The Stono Rebellion ended up as one of the biggest flops in the history of American slave rebellions, and as instructive a tragedy for blacks as it was for whites. Twenty slaves in Stono successfully raided a weapons depot, armed themselves, and left the heads of two hostages on display before brazenly crossing the river to Georgia en route to the promised land of Fort Moosa. Slaying whites and appropriating weapons at every house they passed, the crew gained ecstatic adherents along the way. Slaves joined in with shouts of "Liberty!" as the freedom fighters moved south. But by the end of that first afternoon, government troops descended upon the now dozens of insurgents in an open field where the slaves somewhat prematurely celebrated their freedom with rum-fueled dancing and singing. After the killing stopped, dozens of pikes topped with rebel heads lined the main road at one-mile intervals, courtesy of the King's men. But even that gruesome lesson did not deter the rebels of South Carolina. In short order a follow-up conspiracy would be discovered, with its inevitable punctuation mark of fifty convictions and fifty more executions.

Stono was still fresh news in New York when the mystery fires began. People also remembered the homegrown conspiracy that shook Manhattan back in 1712. Some residents even convinced themselves that blacks had plotted to poison the public water supply the previous year. Many white Manhattanites had abandoned their backyard wells and bought springwater from vendors who trucked it around town in carts. Some residents still felt comfortable sending trustworthy slaves to the spring to fill up buckets and jugs, but others asked themselves: Who precisely could be considered "trustworthy" now?

Investigators broke the Great Negro Plot once they offered money in exchange for information. A sixteen-year-old white indentured servant named Mary Burton volunteered a story about a tavern she knew and the gang of thieves who came together there. Patrons nicknamed the place Oswego, after an English trading post on the shore of Lake Ontario where the riches of Europe were bartered for those of the northern frontier. The fortunes made within Oswego's log palisade were legendary. But none of the exchanges that took place at this faux Oswego—a tavern owned by a middle-aged, married white man named Hughson, located on a disheveled stretch of Manhattan waterfront—were legal. The enterprise at Hughson's Tavern centered around purloined jewels, silver, pieces of eight, firkins of beeswax and butter, coats, and stockings, all lifted by enslaved black New Yorkers from the homes and storehouses of their white owners. Hughson's place comprised the perfect location to stash the goods before transferring them to a fence who would redeem them for cash. No one who patronized the place would ever tell.

But Hughson's Tavern was never just a den of thieves. The establishment's patrons, drawn together over rented tobacco pipes and penny drams of rum, were buoyed by the rush of power that came with succeeding at petty crimes against the white people who ruled their lives. They began to brag about what they would do if they got the chance. About how they would make the chance, damn the consequences.

Joanna could never have guessed at the complexity of this network that united petty crime, arson, and a common determination to overthrow the white elite, but its membership reflected perfectly the increasingly free-flowing heterogeneity of New York. Some of the rebels were black, but disaffected white and Hispanic citizens also took part. One highly visible white participant, nicknamed "the Newfoundland Irish beauty" but also known as Margaret Kerry Sorubiero or simply Peggy, was a red-headed sometime prostitute recently arrived from London. She scandalized local gossips when she took a room in a boardinghouse owned by a free black named Frank. Moving to the tavern brought Peggy into the plot and closer to Caesar, the slave of Vaarck the baker and soon the father of the Newfoundland beauty's baby. She would later bring the baby to the trial to nurse before the judges ordered it removed from the courtroom, and the baby would become an orphan when the province later hanged both her father and mother.

A clutch of Spanish American prisoners who had been picked off a prize ship and sold into slavery also plotted against white New York, as did about three dozen other Irish men and women. Some of these Irish were soldiers who despised the army (including military guards who had been trusted to protect the royal governor) while others vowed revenge after seeing their homeland crushed by the Anglican Church. But most of the insurgent's leaders hailed from Africa. In the New York underworld they had formed gangs: the Geneva Club, whose name came from the most popular gin of the day; the Long Bridge company; and the Fly Boys. Each gang claimed a captain, called a headman.

Cuffee, a slave owned by Adolph Philipse, was a captain. So were Curacao, Dick, York, Bastian, and Prince, each from a different region of Africa. Every gang had a drummer and fiddler for rhythmic inspiration of the troops. Organized ethnic units included the Malagasy from Madagascar (the Philipse family's specialty when importing human chattels), the Igbo from the Niger River, and the Papa from Whydah on the infamous Slave Coast. The leading cell included Gold Coast Akan-speaking Coromantee slaves, known for their warrior training in West Africa. The insurgents, it was said, swore by thunder and lightning to complete their rebellion, and kept poison on their persons to ingest if captured.

With Joanna's attention diverted by scrumptious fashion dolls, the happy clutter of party invitations, and her husband's political trajectory, she would not be attuned to any covert proceedings percolating in the world around her. But even when rumblings occurred directly in front of her lead-white-powdered nose, she was in all likelihood just too consumed with her own affairs to note them. In part, she was willfully blind. She simply could not imagine any reason why a lady of her ilk—pleasant, well-meaning, God-fearing—would be the object of anyone's dislike.

But even protected, privileged Joanna could not escape the events that put the city in a panic, especially once the revolt's leaders were arrested and she bid her husband good-bye every morning as he left to preside over their conspiracy cases. Frederick

Philipse, at the age of forty-three still ensconced on the bench of to the colony's most important court, had an opportunity that surpassed even the Zenger trial to display his superior powers in the name of right.

Chief Justice Daniel Horsmanden forbade Frederick and De Lancey from leaving the city for their summer retreats during the course of the investigation. That was an unprecedented hardship, as the court usually only did a week's work in July. The sheer volume of individual trials was exacerbated by the rolling nature of the investigation: As new sources came forward, further arrests took place, additional suspects came to trial, more convictions came about, and the gallows and stake got a workout. Even as the investigation targeted perpetrators and sent them to trial, new fires flamed.

Taking refuge at Philipsburg Manor might have appealed to Joanna once, but now it seemed less than safe. How could she not fear the place when one of the first men collared was Philipse property? Cuffee had been seen leaving the scene of a Manhattan fire, specifically the fire that consumed a storehouse belonging to Adolph Philipse. In Joanna's view, if Cuffee could conspire under the noses of all of white society in the city, what on earth was taking place at Adolph's Upper Mills enterprise, where two dozen Africans had run of the place, governed only by a part-time overseer and an absentee lord of the manor?

As Frederick declared in his opening remarks to the grand jury: "This crime is of so shocking a nature, that if we have any in this city, who, having been guilty thereof, should escape, who can say he is safe, or tell where it will end?"

Joanna had always considered herself on good terms with the men and women who labored within her household, but now she doubted their loyalty. Her five children—the oldest, Frederick III, all of twenty-one and Margaret, the baby, only eight—had grown rattled by the events of 1741. They would probably have preferred to escape from city and country both, to get on a ship and sail off to Europe—if it were not for the excitement of going to gawk at the suspects being led in chains to the jail on the green in front of City Hall. Living through the Negro Plot would fuel Joanna and Frederick's drive soon after to make their country seat a secure, livable year-round compound.

Before his arrest, Cuffee lived in the city, where as Adolph's house slave he ran errands, served as butler and coachman, minded the condition of his owner's boat, powdered his wig, and helped him button his trousers if those were his orders. Adolph had retired from politics, but as former Speaker of the New York Assembly he maintained his place at the center of New York's intellectual culture. He also still took part in the genteel social life of the colony, although his version tended toward men-only, hard-tippling get-togethers rather than the fancy-dancing affairs staged by his nephew Frederick's vigorous wife. As the city servant of an accomplished unmarried gentleman, Cuffee had greater flexibility in his free time than some of the slaves he knew, but the pressure to snap to another man's demands existed as it would for any other slave. And it tormented him, according to trial testimony.

That Cuffee and the others accused in 1741 had an actual trial by jury reflected less their sympathetic relations with New York slaveholders than a peculiarity of that historical moment. No slave held citizen status, thus none could expect a jury trial by law. The nineteen enslaved men and women accused in the slave plot of 1712 had been prosecuted summarily, by Attorney General May Bickley, in accordance with the stipulations of the slave code that stood in effect at that time—in other words, his role allowed him to unilaterally decide their fate sans judge, jury, or defense team. Some colonists and legal advisers in London saw the punishments of 1712 as distasteful. This time, the justices would protect themselves and dot the politically correct i's by bringing the accused blacks to trial and including citizens as jurors in the process. There would be at least the appearance of fairness. (Unfortunately, all of the city's lawyers had been pressed into service to prosecute the alleged conspirators, and there were none left over to represent the defendants, even if they had had the money to pay them. The uneducated, inexperienced slaves, despised by jury and prosecutors alike, would have to defend themselves.)

Like some others accused of organizing the conspiracy, Cuffee had a reputation as something of a disrespectful thug, a reputation that did much to undermine any chance he had of persuading the jury of his innocence. During the trial it emerged that Cuffee had engineered a cockfight at Adolph's house three years earlier when his owner was absent—an account that unnerved the comfortable with its image of black hustlers penetrating the domestic precincts of rich white merchants. One witness reported that Cuffee "used to say, that a great many people had too much, and others too little; that his old master has a great deal of money, but that in a short time, he should have less, and that he [Cuffee] should have more." A slave named Jack recalled a Sunday at Hughson's Tavern when Cuffee played the fiddle and everyone danced, then downed a bowl of punch and plotted to set fire to the town and kill all the white people.

And yet now, condemned to die, Cuffee lost his tough-guy exterior. On the day of his trial, Cuffee and another slave named Quack faced two felony counts, one for arson and one for conspiracy. Attorney General Richard Bradley made the stakes clear in his directions to the jury of twelve prosperous white New Yorkers, highlighting the urgency of convicting the accused in order that citizens "may sit securely in their own houses and rest quietly in their beds, no one daring to make them afraid." All eleven of the Crown's witnesses had been granted reprieves for testifying and all made statements alleging that the suspects had been heard vowing to start fires, specifically to light up the aging timber storehouse of Cuffee's master. Cuffee had been arrested near the Philipse storehouse on the afternoon it burned, standing to the side in his natty blue coat with its red lining. He was said to be suspiciously callous about pitching in to save his master's property. When people offered him buckets of water, the slave had only "huzza'd, danced, whistled and sung." One witness recalled Cuffee and Quack discussing plans relating to the aftermath of the arson, "killing the gentlemen and taking their wives to themselves." Further playing into hot-button fears of black male predators against white women, one theatrical teenaged indentured servant fingered Cuffee for what she called molestation, referring to his boast that he intended "to have her for a wife."

The two slaves called ten witnesses to testify on their behalf. Adolph Philipse actually offered an alibi for the slave who had allegedly burned his valuable property and plotted to take over the city. Adolph told the court he had assigned Cuffee the task of sawing wood in the morning and then sent him to the waterfront to repair the sail of his sloop at the very moment his building went up in flames. Adolph could not bring himself to salute the man's moral fiber ("As to his character I can say nothing"), but other prominent merchants stood up as character witnesses for Cuffee and Quack. As with past revolts in other places, well-to-do slaveholders defied the expectation that they might jettison their dissident property, instead making an effort to get them declared innocent and released. It was mainly a question of cost-benefit for a merchant like Adolph Philipse, with perhaps a bit of proud denial that the person he "knew" so well would target him.

By the end of the conspiracy proceedings, character witnesses did not matter. The important thing was to make the trials serve, in the words of Justice Frederick Philipse, as "a terror for the evildoer."

As the cold winter gave way to a particularly lush, green spring, New York was determined to eradicate all shreds of the burgeoning rebellion. Authorities had only the confessions of terrified slaves and little hard evidence, but it was no longer a question of plans that had been made or executed, but plans that might be made in the future by "this brutish and bloody species of mankind," in the pungent description of lead prosecutor William Smith.

Cuffee was one of the first conspirators sentenced to death for his often bragged-about but never-accomplished intention to, in the words of Justice Horsmanden, "undertake so vile, so wicked, so monstrous, so execrable and hellish a scheme as to murder and destroy your own masters and benefactors, nay to destroy root and branch all the white people of this place and to lay the whole town in ashes." Execution by burning would take place the following day, May 30.

Half of the city's black male population sat in jail—in the dungeon—while the implicated whites occupied cells aboveground. But as he awaited his sentence, Cuffee knew nothing about the fate of others. Alone in the darkness and damp, he "never mentioned anything . . . but read sometimes and cried much." Led to the Commons in front of the "new" City Hall, which had been constructed at the head of Broad Street in 1700, and asked for the umpteenth time if he cared to admit his guilt and so earn a reprieve, Cuffee reversed all his previous assertions of innocence and perjured himself to stay alive. He offered the names of others to make his statement more valuable to the investigators. Quack followed suit.

All afternoon, a crowd had swelled around the stake and kindling. Suddenly, as Chief Justice Horsmanden scurried off across the green to consult with the governor about a last-minute pardon for the two who had finally admitted their guilt, it looked as if Cuffee and Quack would escape their judgment. But Horsmanden soon returned. Despite the confessions, the executions proceeded as planned: The armed escort could not make its way through the mob to get the men back to the jail.

More to the point, the crowd's desire for resolution, for a proper ending, for a gruesome death, won out over any superficial adherence to rules of fairness. That the lawmen broke the official promise of a reprieve offered a further demonstration of the government's absolute power. And the public executions allowed the government to grant the wish of the God-fearing men and women of the city: to drink in the sight of fire consuming the guilty, from the first smoky crackle to an engulfing vortex of flame.

Of twenty-five defendants found guilty, thirteen were burned alive and twenty-one hanged, including two white men and two white women. Seventy-two others found themselves banished to the Dominican Republic, Curaçao, Madeira, Newfoundland, and other sites where they could no longer threaten the safety of Joanna's New York.

*O*ver the course of the next decade, Joanna and her family tried to put the harrowing events of 1741 behind them. After it was all over, once the memory faded of the corpses laid out to rot on the public commons, the episode most likely even instilled pride in the lady of the manor as she reflected on her husband's role, how Frederick stepped up to staunchly defend the public good under challenging circumstances. Of course it was not easy, from a personal standpoint, for either Frederick or Joanna to witness the death of a person they knew well, no matter how heinous his crime. There was also the danger that Cuffee's execution had infected his Philipsburg peers with bitterness and that they also might rebel. For colonial slaveholders, a death like Cuffee's meant the loss of valuable property as well. But in the view of the Philipses, it had to be. By the start of the 1750s a fragile air of security had returned to white

Manhattan, after it had been cleansed by the obliteration of the rebels, a step that was somberly endorsed by the judges of the court.

In 1750 Uncle Adolph succumbed to tuberculosis, at the age of eighty-five. The old merchant had never seen fit to marry, had never produced an heir. For companionship he relied upon the band of brothers at the Hungarian Club or supped with his step-sister Eva van Cortlandt, who would survive him by a decade, expiring finally at the age of one hundred. Margaret's firstborn out-lived all her siblings, and saw the intimate Dutch-inflected world of her youth give way to a bustling, sophisticated British province.

More surprising than Adolph's choice to forgo wife and children was the fact that at his death the shrewd old merchant-politician had left this world intestate. Perhaps with no family of his own he simply accepted the logical outcome of his passing: the long anticipated reunion of the two halves of Philipsburg, the Upper and the Lower Mills. Adolph's share and Frederick's share would merge into the spectacular fifty-seven-thousand-acre pack-age it had been when Frederick I received the charter for his manor in 1693.

Now Frederick II could luxuriate in his possession of the en-tire parcel. And more. Adolph's nephew would gain all the addi-tional properties Adolph had acquired subsequent to his own bequest from Frederick I. There were numerous city houses and lots, ever increasing in value. Frederick also received 205,000 acres of largely unrealized agricultural promise north of Philipsburg, known as the Highland Patent for its pair of rugged highland ranges that bracketed an exceptionally fertile valley.

Adolph had purchased the Patent, which included all of New York's Putnam County, from the Wappinger Indians in the late 1690s, when he was in his precocious early thirties. King William and Queen Mary soon granted the tract manor status. Adolph's property commenced at the northern lip of Van Cortlandt Manor,

Philipsburg's eighty-six-thousand-acre northern neighbor (founded by Catherine and Jacobus's brother Stephanus). From there his land sprawled upward to meet the southern edge of the next manor along the Hudson, the eighty-five-thousand-acre Rombout Patent, named for owner Francis Rombout, a Dutch-era Belgian fur trader whose career trajectory cut at much the same angle as his peer Margaret Hardenbroeck (with the difference that Rombout ascended to New York mayor, a position for which a she-merchant would never have been eligible). Adolph's Highland Patent bordered the Hudson River on the west and the Connecticut line to the east. Like Philipsburg's, the Patent's mill facilities had begun to attract a growing number of wheat farmers who generated profit for Adolph in the form of rent, more than one thousand pounds annually.

Adolph Philipse likely would have been gratified to see his properties united with those of his nephew. What Adolph surely would not have expected or condoned was Frederick's first act upon the receipt of his inheritance: He decided to unload much of it. Specifically, Frederick sold off much of his uncle's pride, the Upper Mills. Uncle Adolph had built up the value of northern Philipsburg over his half-century's stewardship until it represented one of the fattest cash cows in the family herd. But Frederick did not care for any increased agricultural burden. He had never wanted to be a gentleman farmer, just a gentleman. Just a few months after his uncle's demise, Frederick organized an auction "at the House of the late Adolph Philipse, Esq." to sell a package that included "Four Negro men, viz, a Miller, a Boat-Man, and two Farmers; Three Negro Women; six Negro Boys, and two Girls; Household Goods, and all the Stock, consisting of 40 odd Head of Cattle, 26 Horses, and a Number of Sheep and Hogs, and all the utensils belonging to the said Manour."

Perhaps Frederick felt he needed the cash. His two oldest children had recently married. Joanna, of course, would have played a crucial role as matchmaker, finding her offspring genial mates with suitably deep pockets. Still, love was a costly enterprise, especially for one of the richest families in the colony. In 1745, when he was just twenty-one, Philip Philipse had married Margaret Marston, an English merchant's daughter. Frederick had bestowed upon the newlyweds a cash gift of two thousand pounds, a sum worth roughly half a million dollars in 2006. The couple already had three sons: Adolphus, Frederick, and Nathaniel. Susannah had also exchanged vows, in 1750, with a very respectable catch, the career soldier Beverley Robinson, son of a former governor of Virginia, who had been raised on his family's Hewick plantation in Middlesex County. (In the best self-made tradition, Beverley determined upon entering New York society that his name should henceforth be the presumably more acceptable Beverly. Without the extra *e*.) Joanna's glass-and-silver epergne undoubtedly got a workout when the wedding approached and Beverly's family came to call, all the way from exotic Virginia.

The elite families of the separate English colonies now found it desirable to make forays outside the tribe when locating spouses for their children. A prospective groom's pedigree and fortune mattered the most, and Beverly had strong alliances and kinship ties with Virginia's gentry. He had established himself as a military leader during the short-lived conflict in 1746 known as King George's War, and had upon moving to Manhattan become a mercantile partner of Oliver De Lancey, brother of Frederick II's fellow Zenger judge James De Lancey. Frederick's wedding gift would be the same for his daughters and his sons: He granted Susannah exactly the same sum he awarded her brother.

Any joy Joanna took in her children's nuptials, or in the fact that the family's fortune had more than doubled, would be cut short in 1751. Suddenly that year, her husband of more than three decades died. Frederick was still a relatively young man at fifty-three, so his passing just after his aged uncle was a particular shock to his family.

Unlike his uncle, Frederick did inscribe a will before he died, and one of its foremost provisions captured a central dichotomy of his life. It made sense, from a sentimental standpoint, that this orphan child, raised in the care of his powerful and devout stepgrandmother, Catherine, would be laid out in the family vault beneath the Philipsburg church beside her and his grandfather Frederick. He was buried in the crypt below the stone church overlooking the Pocantico River, according to his wishes. Its Calvinist congregation had continued to thrive as Philipsburg saw an influx of settlers who saw no cause to turn their backs on the faith of their ancestors, unlike the New York City Dutch. Each Sunday, parishioners would accept their communion around the sturdy wooden table protected by Catherine's "five Dutch ells" of damask.

In his will, however, Frederick also ordered the use of four hundred pounds sterling for construction of a new Church of England in Yonkers. The edifice, he decreed, should stand just across the Nepperhan from the Manor Hall—so close as to be immediately visible upon walking out the house's now east-facing grand front door. With this command Frederick all but ensured that his family would no longer consider trekking twenty miles north to the Pocantico to attend Catherine's quaint, plain, Calvinist Philipsburg Church. Frederick clearly expected Frederick, Philip, Susannah, Margaret, and Mary to pledge their allegiance to the Episcopal faith, as he and Joanna had done.

As for the financial aftermath of his passing, Frederick II left no loose ends. The manor went, as was the English custom, to his firstborn son and namesake, Frederick III, who also acquired four New York City houses and half a dozen lots of land, several city storehouses, thirty slaves, the New Jersey salt meadow, a Mamaroneck parcel, and the King's Bridge, which in 1751 still generated a healthy profit.

Frederick II had not totally set aside the more egalitarian inheritance practices of his Dutch ancestors, however, in that he provided handsomely for his other four progeny, awarding each child a house and lot on Manhattan, slaves, and money. To Philip, Frederick left the house and business on Pearl Street where his "uncle, Adolph Philipse, lived and dyed." Stone Street remained a Philipse stronghold: Frederick left Susannah and Beverly the house he himself lived in, on Stone a little east of Fort George; Mary would reside across the street, on the south side of Stone; Margaret would inhabit a house just around the corner from Stone that fronted on lower Broadway; and Frederick III would take the ancestral home on the north corner of Stone and Broadway. Frederick then addressed himself to one of Adolph's most lucrative properties, the bloated tract called the Highland Patent. Each of Frederick II's children beside his firstborn would receive one-quarter of the vast estate, portioned out with the intention of rendering the division flawlessly equal. Finally, having presented both Philip and Susannah with two thousand pounds apiece when they married, their father now pledged the same to each of his two single daughters upon their future nuptials.

Joanna, on the other hand, would receive a simple cash annuity of five hundred pounds. Frederick gave her free use, but not ownership, of the house on Stone Street (the deed would revert upon her death to Susannah), and seven of the upwards of fifty slaves in his possession, identified by name in a codicil to his will as John, Squire, Kofe, Marcy, Creat, Pero, and George. Joanna also received Frederick's coach house in New Street and the garden beside it—not to own, but to use as she wished. (Of the formal gardens at the Manor Hall, the wild acreage that the couple had coaxed into lovely submission, the will said nothing.) And she got wood. Her son Frederick III was to supply her with "50 cords of good walnut" each year, to be conveniently delivered to her Manhattan doorstep.

She received no houses free and clear, no warehouses. No salt meadows. No land at all.

It was simply legal convention by now, this practice whereby a man distributed the accumulated wealth of his forebears among the members of the next generation, with the biggest windfall going to his firstborn son. It had grown common for a relative pittance to go to a wife, even a wife as tied to her husband's political ambitions as Joanna, and even for a man of such unparalleled wealth as Frederick. The new practice went against the ancient principles of Dutch law—against the traditions of the Brockholst family and the Philipse family. A century before Frederick II died, no Hollander in New Netherland would think of departing this earth without providing generously for his spouse, most often leaving her the entire estate or splitting the family holdings evenly between her and their children. But by 1751 it was totally accepted for a husband to convey less to a wife. It was the English way, and no other piece of evidence states as plainly as Frederick's will that the Philipse family, like so many others in the province, was now governed by English custom.

Even so, it is hard to imagine that some tongues did not wag among the tight-knit families of New York's landed gentry. Joanna Philipse had come to her marriage by way of an aristocratic upbringing and had always been cushioned by privilege, accustomed to nothing but the finest in dining, dress, and decor. (How much of her own inheritance remained now that Joanna had reached the age of fifty is an open question.) The value of her five-hundred-pound monetary allowance can only be judged relatively, measured against her customary needs. It was unquestionably a substantial sum, worth about $125,000 in 2006 U.S. dollars. For comparison, as Joanna embarked upon her life as a colonial widow the standard wages for an English housemaid were a token ten pounds per year, a farmhand made just seventeen pounds, and a well-respected clergyman could expect an annual compensation of about one hundred pounds. Joanna's bequest would clearly have provided nicely for the average woman in mid-eighteenth-century New York.

But for a person of her stature, it was like a millionaire reduced to living on nickels. A stipend of five hundred pounds would hardly begin to support a lady, a woman who was still *the* lady of Philipsburg Manor. With a bundle of firewood and some slaves to feed the fire, her husband had ensured that she would stay warm in winter. That was the extent of the creature comforts granted Joanna by Frederick, her devoted husband of thirty-two years.

She would have to scrape to afford the Parisian dressmaker.

Frederick's principal public tribute after his death appeared in the *New-York Gazette,* the paper whose lockstep adherence to the political status quo had stimulated Peter Zenger to produce his own journalistic organ. The item commemorating Frederick Philipse II concluded that "few men [. . .] ever equaled him [in] those obliging and benevolent manners which, at the same time that they attracted the Love of his Inferiors, gained him all the respect and veneration due to his rank and station."

As for the death of Joanna, no one wrote it up in any newspaper. Of a will, we know nothing. Nor has a year for her demise ever been determined, although one genealogist of the family a century later suggested she might have lived until 1765. Legend relates that this always high-spirited horsewoman took a fall while traversing the rough terrain of the Highland Patent and did not survive her injuries.

14

$\mathcal{A}$ Castle on the $\mathcal{H}$eights

$\mathcal{N}$othing so captured the spirit of large-scale eighteenth-century celebrations as an all-day, no-holds-barred, quaff-till-you-drop ox roast. Citizens went all out to observe a major occasion, such as Coronation Day or the King's Birthday. In 1748, for instance, Fort George in New York City marked the birthday of George II by hosting what a contemporary chronicler described as a "snug select company of the *choicest fruits* of the town" to drink "a most extraordinary glass of wine" to His Majesty's health. An overflow ball at a local tavern accommodated all the somewhat less-choice fruits who could not gain admission to the fort. Ending at five o'clock in the morning, the tavern's "gay and numerous assembly" seemed to have been a success; it transpired "without the least incivility offered or offence taken by any one, which is scarce to be said on the like occasions." Perhaps the most outstanding feature of the 1748 celebration was the fact that the city "was illuminated from one end to the other," an undertaking whose expense can only be imagined.

At a New York ox roast—where the ox itself might be bankrolled by the local manor lord—everyone could come and gorge. The party required nothing in the way of proper clothes, either. Fichus and flounces would only be grease sops at one of these parties.

One ox roast in particular presaged the future of New York, America, and the Philipse family. On January 2, 1759, thousands of jubilant celebrants found a perfect nonroyal occasion to consume slabs of barbecued beef. It was a moment many inhabitants of New York thought they would never witness. The change signaled an epochal, creeping retreat of one of the anointed families of the land, and even the eventual downfall of the English monarchy on American soil. The ox was on the spit because the King's Bridge was no more.

For sixty-six years, since 1693, the Philipse family had held a monopoly on passage over Spuyten Duyvil creek with the gated toll bridge, collecting fees for every pedestrian and herd of cattle that made the crossing. Since the Philipse family owned the land on the northern bank of the creek, the possibilities for building a second, no-toll bridge were circumscribed. For years no one had even tried.

Then two locals built a populist bridge and rendered the King's Bridge obsolete.

Benjamin Palmer, something of a hothead (he would later found City Island as a rival port to New York), and Jacob Dyckman, a farmer and a tavern keeper, broke the longtime Philipse monopoly when they constructed their Farmer's Free Bridge. Palmer solicited donations for construction with broadsides that emphatically declared the rights of the common people. Built of dry rubble and crude wooden planking that crossed from Manhattan to the Bronx just outside Philipse Manor lands, the bridge was not fancy. But it cost nothing to cross. Now, the populist bridge officially opened for business. As Palmer later recounted, "there was a fine fat Ox roasted on the green, and thousands from the city and country, partook of the Ox, and rejoiced greatly." While there were any number of local victories of democracy over entrenched privilege in the decades before the American Revolution, this may have been the only one toasted in beef gravy.

But someone lost out in this revolutionary transition. And that unfortunate someone was Mary Philipse, the second of Frederick II and Joanna's three daughters, who had spent her formative years at the house on the Nepperhan, puttering with her mother in the kitchen, weaving love ribbons, and attending fancy-dress parties that filled her house with the beautiful people of the day. As of 1759, the year the King's Bridge became moot, she had been out of the house and married only a year. But she still spent time under its roof—a roof that now belonged to her brother Frederick III, the current lord of Philipsburg Manor—as she alternated sojourns at the Nepperhan with stints at her inherited residence on Stone Street in Manhattan.

Mary was raised to see the King's Bridge as her family's birthright. No longer could she ride blithely across as her mother had done and as Catherine had before her, cantering in front of the pack, tipping her bonneted head to the gatekeeper while all others had to reach into their pockets.

The change had two consequences for the family. One, the Philipse bridge no longer brought in the revenue it had always generated. The second impact was political. Traveling over a bridge for free no longer delineated the difference between a Philipse scion and everyone else. With competition, the power drained out of the toll bridge.

It is not so surprising, then, that three months after the dedication of the Free Bridge an advertisement appeared in *Weyman's New-York Gazette* offering for let "the House, Farm, and Bridge at

Kings-Bridge, in the manor of Phillipsburg, in county of West-Chester." The Philipses were too shrewd to hang on to an investment that brought any trouble. The Manor Hall was another story.

*W*hen guests came to visit the house that Margaret built in Philipsburg at the middle of the eighteenth century, a tour of the ceiling plasterwork was mandatory. Floating above the shiny southeast parlor of the mansion where Mary grew up, curds and furbelows of papier-mâché shimmied across the plaster surface. The once-plain ceiling installed by Joanna Philipse bloomed with sinuous wreaths and flowers woven around two eternal themes: music and the hunt. Shaggy hounds alternated with pelicans, an ancient symbol of Christ. The mother bird, which reputedly sacrificed her life to feed her young with her own blood, mysteriously tucked her outsized beak to her chest. Satyrs caroused on goat hooves and blew on pipes, looking like they were whistling at a girl on the street. A country damsel played a lute. A woman in a cocked hat, perhaps a performer, clasped a songbook. Clusters of phallic acorns popped from fans of cartoonish oak leaves.

Frederick Philipse III and his wife, Elizabeth, the new lord and lady of Philipsburg Manor, had commissioned these plaster fantasies from a master craftsman whose style had been derived from every recent artistic sensation of Paris and London. On either side of the ceiling, two portraits of men in oval frames stared down into the room. The pair represented what the English termed "worthies," and they were an homage to the busts of great men displayed at the Temple of British Worthies, the grand marble structure that overlooked the Worthies River at the famous Stowe estate in England. Completed in 1738, the Temple was the talk of English society both on the Continent and in the colonies. There, at precisely proportioned intervals, the sculpted heads of Isaac Newton and

Alexander Pope were joined by other men of letters, including John Milton and William Shakespeare, and political heroes such as Sir Walter Raleigh, Queen Elizabeth, and Sir Francis Drake. A central pyramid contained a bust of Mercury, there to lead the Worthies to the Elysian Fields.

It would have been difficult to attempt such a grand plan here on the Yonkers ceiling, but the medallions of Pope and Newton stood in for the rest. The two men made for an interesting contrast, floating up there among the maidens and hounds. The brilliant scientist Newton had been a deceptively simple man, a lifelong bachelor who evolved into the most esteemed natural philosopher in Europe before he died in 1727. Pope's genius couldn't have been more different. With a ribald reputation, a height of just four feet six inches tall (from a childhood spinal infection), and a grievous humpback, he could match any scathing remark about his appearance with the lacerating wit of his poems and prose, much of which lampooned the battle between the sexes. Now the two plaster heads, one a theoretical scientist, the other a satirist, observed the passing parade from their place on high, and whatever opinions they formed about the life of the increasingly frivolous Philipse family they thankfully kept to themselves.

The ceiling was one of the first changes Frederick Philipse III made to the Manor Hall when it came into his possession in 1751. The eldest of Frederick II and Joanna's five children, Frederick had known this house all his life. He and his siblings had watched their parents renovate, increasing the light and space of the house, constantly upgrading the life of the family within. At thirty-one, Frederick considered himself more than capable of injecting the place with his own brand of style. The room where the colonial politicos and fashionistas of his parents' generation had gathered already gleamed with luxe surfaces and impressive architectural features. Frederick literally topped the whole production with this rococo powdered-sugar canopy of delicious, British-tinged plasterwork imaginings. He hired a New York–based wood-carver named Henry Hardcastle to remodel the fireplace. Square in the center he placed the head of Diana, goddess of the hunt; she stared ahead with an expression of deathly calm as though nothing that took place in this room could surprise her in the least. A mad array of plants and exotic birds, including a toucan, surrounded the deity. The Philipses doused the finished carving in a coat of sea foam–tinted paint.

These elaborate whimsies took more money and time to indulge in than most colonists could afford. In fact, owning carvings by Hardcastle put the Philipses in an exclusive club. But the young scion Frederick III and his wife, Elizabeth Williams, had all the time and money they needed. When Frederick and Elizabeth married in 1756—she was twenty-four and he was thirty-six—they dedicated themselves to buffing the already sparkling jewel in the Philipsburg Manor diadem, the mansion overlooking the Nepperhan falls that Joanna and Frederick II had recently redone themselves. Elizabeth was the perfect helpmate in this endeavor. Lively and charming, Frederick's bride had been raised in New York, the daughter of a naval officer who was the head of the city's ports.

Previously Elizabeth had been married to the lawyer son of a landed mogul named Anthony Rutgers, who helped shape the geography as well as the economy of Manhattan by draining a swamp at the base of the island that was part of the vast parcel known as the King's Farm, then Trinity Farms—the future Tribeca—so that the land would support his grand pile of a mansion. Later, the Rutgers' estate would become famous as an attraction called Ranelagh Gardens, patterned after the British resort of the same name, and it would draw aristocratic pleasure seekers for decades. Elizabeth lived with her first husband in a mansion fronting the East River, a property the young widow presumably brought with her when she attached herself to the equally moneyed Philipse lineage.

Elizabeth's taste in home decor seemed to volley between the voguish and the staid. She and Frederick mustered forces to create an environment that was outwardly as Palladian, symmetrical, and austere as any mansion going up in England, yet crammed with the excessive rococo features that contemporary fashion followers expected in a residence. And to say they expected them is no exaggeration. Letters and journals of New York's mid-eighteenth-century elite denizens reveal them to be an intensely opinionated bunch when it came to architecture, crafts, clothing, and in fact any material goods. Their antennae regarding finishes, flourishes, and workmanship remained in a state of perpetual arousal. Elizabeth and Frederick III and their fancy friends judged every detail. Hence, the elaborate artistry of a ceiling that would require a telescope for proper viewing. And the perfect Diana afloat in sea foam.

Of course they would have no qualms about spending money on home renovations. They had too much not to spend. With no banks, they had to secrete all that lucre someplace. And once they started renovating, how could they stop? It was now really a question of adding embellishments. Frederick's great-grandparents, parents, and parents had built, then built on. His job, with Elizabeth, would be to burnish. She would intersperse her decorating efforts with bearing eleven children (nine would survive to adulthood), and supervising dozens of servants and slaves.

Frederick had the time to devote to burnishing because his involvement with the manor was more like the approach his father had taken than that of his great-uncle Adolph or his great-grandfather Frederick I. Philipsburg Manor had become principally a pleasure palace inhabited by Frederick and Elizabeth, their growing brood of children, and Frederick's sister Mary, with occasional visits by sibling Susannah and her husband Beverly. Records suggest that Frederick III had positive relationships with his tenants—he was famous for his policy of never raising the rent on a property during the lifetime of a tenant. He embraced the responsibilities of rent collection, but that was about as far as he wanted to take it.

Embraced might even be too strong a term. Frederick III approached the duties of a manor lord with an unprecedented languid, effete mildness of purpose. He didn't like to get his hands dirty, so he left that to others. He trained with the Westchester militia as a young man, but wore the rank that he earned—colonel—more as a decoration than any kind of statement of accountability. Frederick carried the "colonel" moniker throughout his life, though he served zero time in any active military capacity. In any case, if he had decided to use his commission, the physical demands likely would have laid him low. Even as a young man, the heir to all Philipsburg stretched the seams of his breeches. As he reached middle age he traveled alone in his exquisite "horse and chair" because his girth kept him from squeezing into a standard carriage alongside his charming wife.

Frederick did have one responsibility besides living the good life. Along with his inheritance he gained the assignment to oversee the completion of the Episcopal church adjacent to the Manor Hall. It was a project the Anglican convert's eldest embraced, as was proper for a firstborn Philipse son. For the first year after her father's death, Mary could chart the progress of the church through the second-floor Manor Hall windows as it rose a quarter of a mile away. In 1752 St. John's Episcopal Church was complete.

Frederick did not wait until the church was done to embark on a few of his own projects, not all of them as fanciful as papier-mâché ceiling decorations. Frederick III cared even less than his gentleman-politician father for the Upper Mills. He found immersion in the details of the mill operation tedious, so he immediately leased the property and put up for sale the remainder of its two dozen slaves. To cut bait on that property was not something Frederick III's great-uncle Adolph or great-grandfather Frederick I would ever have anticipated, much less approved. But the property, and the decisions, belonged to the new lord of the manor. After advertising for a year, Frederick found his tenant.

Mary's brother held political positions, just as their father had, but whereas Frederick II was a Supreme Court justice of some prominence, earning both admiration and approbation for his decisions, Frederick III managed to attend every meeting of the New York Assembly (he had been awarded a position out of deference to his status and wealth) without ever causing a ripple in the discourse. Over the course of twenty-five years, he avoided debates, never volunteered his opinion on any matter at hand, and made himself scarce in committees that dealt with finance or defense. As war descended upon the province of New York in the late 1760s, Frederick asserted only that "he uniformly opposed every measure . . . inconsistent with the rights of crown and parliament." He did interrupt assembly proceedings at one session in 1773, on the eve of the Patriot-galvanizing Tea Act, to introduce an urgent bill of his own "to prevent the killing and destroying of partridges and quails on the manor of Philipsborough." Aside from that oddly timed gambit, he remained in the prewar years a cipher in his beliefs, apart from his belief in the good life.

In the 1750s and 1760s, the pursuit of leisure at Philipsburg took on the status of both art and profession. One observer praised the Manor Hall as the "hospitable center of all that was best in the life of the province." Parties took precedence as they had since Joanna played hostess (she still took part, but left the marzipan sculpting to her daughter-in-law). The Hudson represented a vast backyard; people often reached the house by sloop. In winter the family and guests took sleigh rides on the rock-hard Hudson from the Yonkers wharf all the way down to Manhattan Island. Frederick III created a deer park to the west of the mansion on grounds that sloped down to the river's edge, and enclosed the park with a tall picket fence so the deer would not decimate the estate's tobacco plants.

The Manor Hall's formal gardens dazzled visitors, who toured the acres of terraced roses and other cultivated blooms with the colonel as their guide. Frederick's contemporaries noted his particular interest in the art of landscaping: This "well tempered, amiable man," recounted John Jay, had a penchant for "gardening, planting, etc., and employed much time and money that way." He shared his devotion with intellectuals of the era; he just had more time to devote to the hobby. At the Manor Hall, a manicured lawn stretched from the east facade to the Albany Post Road. Gardeners from Europe reconceptualized the longtime beds of rare plants, dug new walks, and put in original ornaments. Guests ascended a staircase that stood to the north of the house and entered formal pathways that made a circuit past botanical delights. The colonel's focus on his gardens seemed to preclude his focus on anything else, with the possible exception of dining well.

From the time of her father's death until the day she married, Mary resided with Frederick and Elizabeth in the house she had always known, watching it take new form under the eye of a woman who naturally did things differently than she would. She saw her sister-in-law go about the house, furiously decorating, overseeing food preparation, organizing balls and teas, and managing her ever-expanding pack of children. Mary, though, could stay at the house, wander the premises, ride through the countryside, and paint landscapes. Already in her midtwenties and fast approaching spinsterhood, she could escape the countryside and venture south to New York City to visit Susannah. With no husband and no children, she was responsible only to herself.

Not that Mary had no ambitions. Among other dreams for her future, Mary hoped to one day construct her own house. She would deploy all that she had learned as a student of this fine structure on the Nepperhan—its walls of stone and brick, solid floors, rows of generous windows, choice location atop a breezy hill—to create a new home for her future husband and their children. One requirement: It had to overlook a river. The one river. The Hudson. She probably would not go for the ornate touches her sister-in-law favored, like the white-on-white fantasy of a fantasy that floated above the west parlor. Why add artifice to a room that was deliberately positioned to let in the natural beauty of all outside? But one thing was certain about the papier-mâché parlor.

It made the perfect stage for a wedding.

*F*or the cultivated young ladies who paddled in the effervescent social froth of 1750s New York, a fitted uniform jacket with epaulets and saber constituted a potent aphrodisiac. Like the militia-wowed sisters Jane Austen would portray in *Pride and Prejudice* a half century later, the young socialites of the Hudson Valley relied upon a cadre of soldiers to fill up their dance cards, in this case the officers who rotated through New York during the French and Indian War, which dominated the life of the colonies from 1754 to 1763.

In some ways the conflict represented business as usual for the superpowers involved, England and France, and for the men who made their careers as warriors. The conflict followed three wars—King William's War between 1689 and 1697, Queen Anne's War from 1702 to 1714, and King George's War from 1744 to 1748—waged to determine who would win the great territorial prize of North America. None of these wars had proved decisive. This time, the stakes were higher. The French had expanded their New World empire to include the four bustling cities of Montreal, Detroit, New Orleans, and Quebec—a region rich with missions, fortresses, and trading posts. British colonists peered across the Appalachian range and coveted the rich lands of the Mississippi and St. Lawrence watersheds, territories France controlled. Skirmishes began on the remote frontier of the Ohio Valley well before the two nations officially declared war in 1756, when Lieutenant Governor Robert Dinwiddie of Virginia persisted in granting colonists tracts of land that French colonists insisted were theirs. During the war, both the English and the French showered pounds of sterling on various Indian tribes in an effort to buy their loyalty, and natives were pressed into allying with both European powers as well as hitting the warpath to settle long-standing grudges of their own. Fighting ceased with the surrender of the French army in 1760, but the conflict would not officially conclude until the Treaty of Paris on February 10, 1763. The treaty awarded all of North America east of the Mississippi to the British (with the exception of New Orleans, which France transferred to Spain as compensation for Spain's surrender of Florida to the British). England had won this round.

But the French and Indian War served another function in addition to putting most of North America in British hands. In the late 1750s no one could have predicted the harrowing, heroic, world-altering episode that lay just around the corner in British America. Yet the French and Indian War added to the growing tension within the British colonies that would only resolve itself with a revolution. The recognizable players of the Revolutionary War, the Whigs and Rebels, the Tories and Patriots, were not yet in evidence. Still, there were hints of what was to come. The imperial management of the war galled the citizens of the American provinces, most famously when English prime minister William Pitt decided at a low point in the fighting in 1757 to force colonists to provide equipment and supplies, shelter, and manpower for the war effort. Pitt's arrogance—"I am sure I can save this country, and nobody else can," he pronounced in 1756—epitomized the attitude of English leaders as they grew persuaded that London should keep a tighter reign on the colonies, including controlling settlement on those enticing formerly French lands out west. The Crown even posted a ten-thousand-man army along the ridge of the Allegheny Mountains to keep settlers from streaming westward as they desired. It was the equivalent of the Philipses' King's Bridge in what was now a Free Bridge society.

The French and Indian War served as a military preparatory school for a generation of Revolutionary leaders. No matter what their allegiance in the conflict to come, these men gained intimate knowledge of the craft of warfare and the terrain on which they would spill more blood in only ten years' time. It was, in effect, a feeder system for the Revolution, one distinguished by the irony that its cadre of soldiers who fought so fiercely together against the French on behalf of England soon would split irrevocably, with some declaring themselves as Patriots and the rest for the Crown.

But in the Manhattan parlors of the 1750s, it seemed that a major purpose of the war was to serve up waltz partners to the debutantes of New York. For a young woman such as Mary Philipse, strong martial credentials were a desirable accessory for a gentleman, but nuances of ideology mattered less than how an officer filled out his uniform.

Some of Mary's suitors had already won distinction with their service. One was the Virginia-born military prodigy George Washington. When Washington met Mary in 1756 he was twenty-four years old and in the middle of an arduous five-hundred-mile horseback expedition. He had stopped off in Manhattan en route from Fort Cumberland in Maryland to Boston to confer with General Shirley, commander in chief of His Majesty's forces in America, and stayed for a week and a half in the Stone Street household of Susannah and Beverly, who had known Washington as a schoolmate in Virginia. Mary happened to be visiting her sister.

At the time of this visit, his first to New York, Washington had already received acclaim for his performance during one of the initial engagements of the still-undeclared war, the attack on Fort Duquesne in Ohio. From 1753 to 1758 he served as commander of the Virginia Regiment, defending the frontier against the

French. In July 1755 he led his troops at the Battle of Monongahela, an entanglement that came to be known as Braddock's Defeat in honor of William Braddock, the distinguished British
general who led the charge and who met his end there. Washington, then twenty-three, had just spent ten days before the battle
confined to his camp bed with a violent illness. He managed to
bring his thirteen hundred well-armed troops up against just three
hundred French and Indian soldiers—who nonetheless trounced
the British army when its soldiers turned tail and ran. Braddock
received a mortal wound after leading men on fields of battle for
forty-five years. Washington, surviving the rout with four bullets
through his coat and two horses shot out from under him, took
charge and became a hero.

Mary had to admire Washington. She would be pleased to converse with such a distinguished soldier. But compared with some
of the other eligible men she encountered in the same parlor, Washington came off less well. It would have been hard for Mary not to
notice the pits in the skin of his face, first of all. At the age of nineteen, in 1751, he had traveled to Barbados with his half brother
Lawrence in the hope that the climate there would ease Lawrence's
tuberculosis. While Lawrence's condition saw little change, Washington was "strongly attacked with the small Pox," according to a
notation in his personal diary. The young officer may have been a
hero, but his scarred complexion and stiff, awkward demeanor
were hardly an invitation to lovemaking.

Mary had no lack of admirers. A veritable laundry list of their
names appeared in the form of a comic effort produced by a soldier who signed himself "Timothy Scandal Adjutant." This satiri-
cal roll of writ, called "A Return of the State of Capt. Polly
Phillips's Dependant Company, with the Kill'd Wounded, Deserted, and Discharg'd &c, during the Campaigns 1756 & 1756
[*sic*]," noted the status of thirty-nine officers who had a hankering
for "Polly"—Mary's nickname. General Webb, observed Timothy,

was "wounded but tis hop'd may serve." Captain McAdam was "Desperate." Captain De Lancey was "Not accepted." Mr. Gordon, an engineer, "made approaches but at too great a distance." As lengthy as the roster appeared, Scandal noted, it must be incomplete, owing to the overwhelming numbers of Mary's fan club. He urged any "Gentlemen who have any great Demands on the Captain . . . to apply immediately, as we have great reason to imagine the Company will soon be broke."

One Captain Roger Morris, it seemed, experienced suffering more grievous than any of the other officers described: Morris, according to Scandal's spoof, had been "Shot Thro the Heart." The son of a successful English architect, Morris held an officer's commission that his family purchased for one thousand pounds in 1745, when he was eighteen. In 1755 he served alongside Washington at Monongahela, where he was wounded with an injury that put a crimp in his fighting career. Afterward he became brigade major on the staff of Major General Daniel Webb, a commander known for his overall timidity and ineptitude, who had stopped off in New York City in 1756 on his way to the northern front. There, Morris met Mary.

Battlefield accounts fascinated young women such as Mary. She was rather sheltered in her privileged world of imported clothes, jewelry, chocolate, coaches, and tea. It had been a long time since the adventures of Mary's great-grandmother, who sailed to far-off ports and traded with exotic people. Mary had been raised on the site of a Lenape village that was now long gone; even the ghostly midden that told of past feasts had been covered over. The only natives Mary met were the men and women who came to Philipsburg every few years to pay tribute at the graves of their ancestors and sell their intricately woven baskets.

But these soldiers, Washington and Morris and the rest, had come up against real Indians on the Ohio frontier. What were they like? Did they go naked? Did they smash their enemies with clubs or were they using flintlocks? She had heard of scalping, that practice whose ubiquity in the French and Indian War spread from natives to soldiers both English and French. She enjoyed being regaled with the story of a victory in the wilderness. And it was exciting to listen to the horrors of a rout. Morris and Washington had been in the biggest rout in recent history, under vicious but fascinating conditions.

In a July 1757 letter, a friend of Washington's, fellow officer Joseph Chew, goaded the rising-star Virginian about his competition for the pre-Revolutionary It girl of greater New York. In it he said, "I often had the Pleasure of Breakfasting with the Charming Polly, Roger Morris was there (dont be startled) but not always, you know him he is a Ladys man, always something to say." The town, he continued, "talk't of it as a sure & settled Affair," but "I can't say I think so and that I much doubt it." He promised in closing "to set out tomorrow for New York where I will not be wanting to let Miss Polly know the sincere Regard a Friend of mine has for her" and, in the emotional style of the day, "I am sure if she had my Eyes to see thro she would Prefer him to all others."

Mary did not prefer the pitted, taciturn swashbuckler Washington. To Mary, the young soldier had something of the foreigner about him, coming from the southern stronghold of Virginia, so far from New York's Dutch-steeped Hudson Valley, though his planter parents' families had been in America since 1657, almost exactly as long as Mary's own forebears, and though Mary's brother-in-law Beverly also hailed from Virginia. Roger Morris, on the other hand, came from a more purely British background than Washington. At the same time, he seems to have presented himself with an amiability, even a vulnerability, that might have appealed to a headstrong heiress who preferred a helpmate to a head of household.

Roger's "Englishness" was enhanced by the educated aesthetic that came of his upbringing as an architect's son. He surely had "something to say" on any number of topics that would engage Mary's interest. He might even bring some design ideas to the continual redevelopment of the family's manor hall in Yonkers. Or any property they might own together in the future. And given her inheritance they might well own more than a few. Roger's mannered sensitivity shone through in a surviving unsigned portrait of the captain in full uniform—all big moist eyes, Clara Bow mouth, and ever so slightly receding chin. He certainly seemed different from the other officers Mary met in the Robinsons' parlor on Stone Street in New York City, brash cardinals in medal-encrusted waistcoats and form-fitting breeches. Like many Englishmen of his generation and class, Roger Morris probably drifted into military service for lack of a better idea. And Mary Philipse, fortunately, came up with one: The middling fighter should instead become a first-class husband.

The marriage of Captain Roger Morris and Mary Philipse was a high point of the 1758 winter social season, at least if you were a member of one of the few landowning, power-wielding families of New York. Joanna Philipse was still alive for the event; five days before the ceremony she witnessed a prenuptial contract between the fiancés that spelled out the egalitarian inheritance terms Mary and Roger preferred. (The Dutch institution persisted among Hudson Valley families until late in the century.) There might have been a tinge of sadness that day because Mary's younger sister Margaret was not there to see her wed. Always delicate, she had gone to her grave six years earlier, at the age of nineteen.

The wedding party's members arrived at the Manor Hall by sleigh, including Van Cortlandts and De Lanceys as bridesmaids and a dashing young bachelor named Heathcote, who was lord of the neighboring Manor of Scarsdale, as one of the groomsmen. The Reverend Henry Barclay, rector of Trinity Church in New York, arrived after a bracing drive of sixteen miles. The event was a picture of elite British society. The high Anglican ceremony took place in the southeast drawing room of the Manor Hall, beneath the creamy rococo pageant of satyrs and pelicans. A crimson canopy hung over the proceedings, though, interrupting guests' view of the sober visages of Newton and Pope. Instead the assembly was treated to a large gilt rendition of the family crest.

The Philipse crest, a "crowned demi lion issuing from a coronet," formed something of a late grace note for the family's lineage. The ancient pedigree declared by such a crest was a convenient fiction, dreamed up at some recent juncture, probably by Mary's father, Frederick II, to counter the *nouveau* in the family's *riche,* in the venerable tradition of English merchants and country gentry raised abruptly by the Crown's favor to the peerage. The first display of the Philipse coat of arms appears on a silver teapot of Dutch-American design, produced in New York by Jacobus Boelen around 1729. It would not be hard for Boelen or another silversmith to create a crest according to the whim of his client.

For this occasion Frederick III, who gave away the bride, took license beyond displaying the family crest. His attire is said to have included the suitably antique "gold chain and jeweled badge of the ancestral office of Keeper of the Deer Forests of Bohemia," an artifact that could have been found in a Cracker Jack box for all its authenticity. Such historical fakery was honored by Philipse's fellow New Yorkers, a group equally invested in putting on airs. A banquet followed the ceremony (in the elaborate tradition of Mary's mother, Joanna).

At the end of Mary's wedding day, an event transpired that might have foretold the Philipse family's future, if anyone knew how to read the signs. According to family histories, the feast was cut short by the appearance of a mysterious Indian—one who must have come a distance, since the Lenapes of Philipsburg had long since been driven off—who stood wrapped in a scarlet blanket at the threshold of the banquet room. The latecomer uttered just one prescient sentence: "Your possessions shall pass from you when the Eagle shall despoil the Lion of his mane."

*R*oger Morris retired his military commission six years after marrying Mary, in 1764, a step that Mary might have considered overdue. In the first years of their life together Roger had been gone quite a bit as he continued to serve in the French and Indian War. He had acquitted himself well in Nova Scotia against the Indians, been attached to the Louisburg grenadiers in Wolfe's expedition against Quebec, and, as lieutenant colonel of the Forty-seventh Regiment, had commanded the third battery in the expedition against Montreal, under General James Murray. Advancement in the British army earned Morris the respect of his peers, if not the adulation of his generation, like his fellow officer Washington. He spent enough time garrisoned in New York so that the couple could soon begin a family. Their eldest, Joanna, known as Nancy, was born in 1761, and Amherst followed two years later.

But making a lifelong career of soldiering was less than optimal for a man who walked in Philipse shoes. In Mary Philipse's world, individuals with something to lose did not engage in military activities that risked their well-endowed necks. Also, as a married Englishwoman in the colony of New York Mary could not legally own land; she must relinquish all she owned to her husband, at least in name. This stricture made it difficult for Mary to do business without Roger Morris around to produce a signature when necessary. The men who became Philipses-by-marriage became landowners by proxy. Any spouse to Mary needed considerable freedom from other commitments to help manage her properties, assist her in making investments, and put in the obligatory status-upholding appearance with political bodies. Beginning in 1765, Morris served on the legislative council of the colony.

Mary wintered with her husband at her inherited Stone Street house. But the structure never seemed quite large enough, and in 1765, after Mary delivered her third child, Maria, the Morrises purchased a second lot and house adjacent to the Stone Street property. They immediately tore down the timeworn building and in its place constructed a square Georgian house of modern red brick, which had four large rooms upstairs stacked above four below. Its cachet was guaranteed by the architect they hired to draw up the plans, a hot local talent named John Edward Pryor. Just across the street Frederick III and Elizabeth owned an equally substantial town house, and Susannah and Beverly lived nearby. The Philipse sisters, their husbands, and their growing families escaped to the pastoral Manor Hall each summer. Susannah and Beverly had built their own grand country estate, named Beverly, in 1750. It lay north of Yonkers in the village of Garrison, at a river landing nearly opposite West Point. The sprawling wooden structure had already become a landmark.

The first fifteen years of Mary's marriage would be the last peaceful time before the Revolution descended upon the colonies. Certain events of the mid-1760s stood out as flash points for the conflict to come. The British had won the war but lost the peace, at least insofar as the debt they bore when the peace was finally signed. Soon, Parliament passed a quartet of regulations that incited the ire of American colonists. There was the Sugar Act of 1764, which taxed molasses and sugar and put a stranglehold on the rum business; then the Stamp Act of 1765, which imposed a levy on newspapers, legal documents, books, and even playing cards; the Quartering Act of 1765, which mandated that colonists must open their homes to board British troops; and the Townshend Acts, which taxed not only tea but also paint, paper, lead, and glass. It seemed perfectly logical to the Crown that those who lived in the colonies should shoulder some of the burden of their own protection afforded by victory in the French and Indian War. But slapping an excise on everyday trade struck at the heart of life in the colonies and became a daily irritant, an ongoing dramatization of overweening English control. The galvanizing Boston Tea Party would not take place for another ten years, but many colonists had begun to question the very presumption of Parliament to impose taxes at all.

For the rich women of New York, though, the 1760s were years of peace and growing opulence. By this time colonists had witnessed such mind-boggling new technology as lightning rods and Franklin stoves. The price of "flying machines," horse-drawn carriages, had dropped enough so that every well-to-do family could have its pick of rig and paint. Education became a sign of status. The sons of New York's aristocracy matriculated at King's College, the precursor to Columbia University, which opened in 1754 in a schoolhouse adjoining Trinity Church and offered academic instruction under Anglican auspices (but with a mandate of religious liberty, a sign of the changing times). As late as 1748, cultural critic Cadwallader Colden bemoaned the intellectual desert of New York, stating that although New York was the richest of the colonies, it had "less care to propagate Knowledge or Learning" than any other province. "The only principle of Life propagated among the young People" of New York, he noted, "is to get Money, and Men are only esteemed according to what they are worth—That is, the Money they are possessed of."

After acquiring money, the primary goal of at least some New Yorkers seems to have been to acquire the accoutrements of style. The education of affluent girls, often at exclusive boarding schools, emphasized deportment. Paris fashion's siren song drew more followers than ever. The "polonaise," the new-style gown imported from France that featured a full, ankle-baring skirt, filled the closets of Mary and her peers. (Not true closets, since closets were not yet used for clothes storage, but standing Chippendale-inspired wardrobes that had slimmed down and sprouted froufrou since the stout Dutch *kasten* of yesteryear.) Fashion grew airy, light. Instead of extending the sides of their dresses with panniers, ladies accentuated their hindquarters by poufing up the yardage of their gowns in back with ribbons that pulled the fabric up like a balloon shade. The resulting profile was nothing but an S-curve, with jutting bosom, corset-shrunk midriff, and bulging buttocks. Cartoonists lampooned the look, but women of Mary's ilk embraced it.

Ladies' hairstyles from the 1760s through the 1780s became more elaborate and full of interesting details and arresting incongruencies. An upswept hairdo, sometimes called a high roll—elevated, eventually, to a height of a yard—might incorporate pearls, then jewel-studded pins, then peacock feathers, then locks of hair appropriated from a woman's father or lover. Accessorizing the hair was an innovation of French Queen Marie Antoinette, whose fashion influence after she married Louis XVI in 1770 exploded over the course of America's Revolutionary era. For evening dress, some women adopted the white powdered finish of men's wigs, while men returned to a natural chestnut color for their perukes, or even let their shaved pates grow out and tossed their fake hair completely. Some women enjoyed so much leisure they spent a full day in preparation for the night to come, and on occasion a male admirer entered the dressing sanctuary for hours of chat while a socialite's hairdresser worked his magic on the "head," as such a coif was called. To balance the height of her hair, a chic woman increased her bosom, or at least the appearance of it, by artfully inserting sheer, puffy kerchiefs in her low-cut bodice, a fad termed the "pigeon."

*T*he Philipse-Morris household moved easily among its residences. Sometimes the family took its carriage north on Kingsbridge Road, up Break-Neck Hill, past a farm high above the river that had abundant vegetable gardens and an orchard of thriving quince trees. From the crest of the hill the site commanded views in all directions: down to New York Harbor, Staten Island, the Hudson River, the Harlem River, Hell Gate, and even the Long Island Sound. During the ride Mary and Roger liked to talk about where they would situate a country home of their own.

In 1765, the same year they built the Stone Street residence, Mary and Roger bought the property at the crest of Break-Neck Hill from the farming couple that owned it. They laid a foundation beside the quince trees, and planned to design their dream house in the style of a Palladian villa. They called the place Mount Morris. Roger brought his family's architectural acumen to the plans for of the structure, while Mary brought the panache she'd honed living with the perpetual expansion and ornamentation of the Manor Hall. Though a professional architect undoubtedly served on the project, his identity has never been definitively established. One contender, however, is Theophilus Hardenbrook, a local surveyor who also taught architecture, who in his promotional materials boasted his familiarity with various sophisticated styles, and who may well have been a distant relation of Mary's through her Hardenbroeck great-grandmother.

Mount Morris was huge, eighty-five hundred square feet, its hulking frame covered by a coat of glossy, stylish, expensive white paint, formulated from hundreds of pounds of pure white lead imported from England and dozens of gallons of well-boiled linseed oil. Although the structure of a dwelling mattered, house paint was itself a thing of stylistic import, and white pigment the supreme commodity in the house trade of the late 1700s. Mary would have no interest in the tired "stone" colors of brown and gray, or basic red, ubiquitous because it was cheap. Her bright white castle on the heights beamed out style, sophistication, and wealth.

The south-facing front of the house had a double-height portico with a triangular pediment and four colossal Tuscan columns;

the grand front door opened into a spacious central hall. The design did not stint on technical improvements, such as the brick-lined outer walls that protected against summer heat and autumn damp. Roger and Mary took an interest in faux finishes, using flush boarding across the south and west elevations, and adding simulated quoins at the building's right angles to create a semblance of stone. The back, to be seen primarily by servants and children, was left as simple clapboard. (That luscious white pulled it all together, though.)

The feature that truly distinguished the house from an architectural standpoint was the shape of the downstairs drawing room, or the "withdrawing," in the nomenclature of the eighteenth century. A chamber of octagonal shape, the drawing room occupied its own wing, with another octagon atop it, this one divided into two bedrooms and a spacious closet. Scholars deem the design of Mount Morris to be the first time an octagonal form enclosed living space in the history of the colonies.

The villa took over three years to complete. It contained nineteen rooms (counting the great halls on each floor), including seven bedchambers, a surfeit for the unusually modest size of the family. In 1770 the Morris clan expanded further to include Mary's last baby, Henry Gage. Then she was done, breaking with tradition by limiting her offspring to four. Now everyone under Mary's roof could have their own room, with one to spare for a visiting sister. A guest could get lost roaming those bedrooms and the withdrawing room, formal parlor, dining room, basement, kitchen, pantry, root cellar, wine cellar, dairy, laundry, and finished garret.

But in roaming, a visitor would never come upon rococo carvings such as those commissioned for the parlors at Philipse Manor Hall. Such interior flourishes had no place in Mary's interior design, which stressed oversized, simple, even austere moldings in every room. The most ornate embellishment would be the house's wallpaper, which in late-colonial New York fashion might be flocked in a damask pattern or covered with large-scale atmospheric scenes such as the Roman ruins or the falls of Niagara.

Outside, stables and a coach house improved the grounds. The couple also "laid out a great deal in stone fence," in Roger's accounting. The bill for the project came to thirty-five hundred pounds, a small fortune even for the Philipse-Morris family. But it was worth it. Though the couple built the mansion primarily as a summer retreat, it fell conveniently close to two other principal homes and places of business, being almost exactly equidistant from Stone Street and Philipsburg (the Highland Patent was more remote, of course). Mount Morris was but a bump or two down the road to the King's Bridge. And yet their home felt private because it stood on a vast chunk of land, over one hundred acres that ran straight across the top of Manhattan, all the way from the Hudson to the Harlem River.

Mary and Roger were at the top of the world.

Yet one problem close to home seemed impossible to resolve.

The heftiest percentage of the couple's income now derived from Mary's portion of the 240-square-mile Highland Patent. A 1753 itemization of the entire property's features cited "five hundred messuages [dwellings with outbuildings and land], twenty mills, twenty dove houses, five hundred gardens, two hundred thousand acres of land, one hundred thousand acres of pasture, two hundred thousand acres of wood, thirty thousand acres of Marsh, ten thousand acres of land covered with water and common of pasture for all cattle." Though Mary owned just one-third of this tract, it clearly overshadowed other business ventures she and Roger indulged in, such as the downtown Manhattan establishment called Freemason's Arms that was crowded with twelve fireplaces and two ballrooms. An advertisement in the *Mercury* of May 2, 1768, offered to lease the "noted tavern" belonging to "the Hon. Roger Morris Esq.," enumerating its "great run of business for many years past."

By the time Mary married Roger, Beverly Robinson's eponymous and lavish mansion had become the official manor hall for the entire Patent. Though Beverly took the lead in managing Philip and Mary's estates as well as Susannah's, Mary and Roger also participated. The couple built a log house, which Mary made her base when she arrived to collect her tenants' rent. She gained a reputation for the miles she clocked as she made her rounds on horseback, most likely outfitted in the au courant long-skirted riding habit called the Joseph, which was only chic when styled of fabric the color of fresh-shucked peas. Local memory has it that when Mary was in the neighborhood, residents crowded the upper loft of a local gristmill in the hope of catching a peek at her.

Management of the Patent in the 1760s offered Mary Philipse more cause for worry in the day to day than any tax on consumer goods imposed by England. In actuality, running the estate meant dealing with a legacy of controversy, the only instance of reality chafing against the comfortable Philipse lifestyle of the 1760s. When Mary and her siblings inherited the 205,000-acre parcel in the early 1750s, just a hundred or so Europeans had settled only a scant portion of its acreage. Members of the Wappinger Confederacy were the tract's other residents. Unlike other natives in the Westchester region, these Indians had not fled west after a century of strife and sickness. Not that there were many. Scholars peg the native population on both sides of the Hudson River at only three hundred individuals in 1774. But the Wappingers who remained asserted their presence.

At the onset of the French and Indian War, the Crown had recruited the Wappinger troops of the Patent to fight the French in Canada. Colonial authorities then relocated the remaining women,

children, and the elderly, ostensibly for their protection, to a Chris
tian Indian mission settlement at Stockbridge, Massachusetts, for the
duration of the war. The youngest generation of adult Philipses—a
generation for whom this Patent land represented their major holdings and principal source of income—saw a business opportunity
with the natives' absence. (Adolph, by contrast, had cultivated the
moneymaking potential of the Upper Mills, and had left the Patent
basically alone.) The Patent's value depended upon leasing property
to rent-paying tenants. This seemed the perfect moment to bring
in new settlers to inhabit the land of nonpaying inhabitants, namely
the natives who had left to serve as soldiers for the Crown.

There was only one problem. Even though their numbers had
dwindled, the Wappingers insisted that they were the true proprietors of the land. They had never accepted the authenticity of
Adolph Philipse's original claim of 1697, insisting his official patent
for the land was invalid because they had not sold him the entirety
of the parcel. They contended that they legitimately owned the
easternmost three-quarters of what would become Putnam County.
The Indians had never stopped acting as the original landlords of
the Patent, and had sold outright or leased land to many settlers
(mostly New Englanders), in large part to establish their own claim
on the tract.

At the conclusion of the war, the Wappinger soldiers returned
to see their lands occupied by colonists with fresh, Philipse-granted
leases. In order to settle paying tenants on the Patent, the Philipses
had seized properties and ejected tenants who held Indian land
titles, burning some of them out of their farms. Throughout the
Hudson Valley, where similar land disputes exploded, civil authorities brought in troops to control protesters. Now the Indians were
not the only ones aggrieved. In 1765 and 1766 displaced tenants ri-
oted against the Philipse family. Mobs forced Philipse-installed
Highland Patent tenants off their lands and burned their barns. The

New York government finally suppressed the rebellion, but not before Beverly Robinson found it necessary to flee for his life to New York City.

In the 1760s, as Mary Philipse reveled in her modern house at the peak of New York society, the Wappinger sachem (and, later, Revolutionary War hero) Daniel Nimham brought repeated claims that Adolph Philipse had lied about his land deals with the Indians. Mary would have kept a close eye on Nimham's campaign to persuade the colonial powers-that-be to compensate the tribe for the hunting grounds it had lost to new settlers and their sprawling fields, pastures, and orchards. Nimham even traveled to England to petition the king for a return of tribal lands. It was a high-profile dispute, one covered extensively in the newspapers and hashed over by manorial society.

After a favorable royal hearing, the unstoppable sachem approached New York's Governor Henry Moore and the Common Council to file yet another claim to recover 204,800 acres of Wappinger land from Philip Philipse, Roger Morris, and Beverly Robinson. The Philipse men employed a team of fifteen lawyers in their defense, while the Indians by law could not even testify on their own behalf. (One settler, Samuel Monroe, attempted to assist the Indians but was jailed to silence his testimony.) Council member Roger Morris was permitted to hear the case despite the obvious conflict of interest, though he refrained from signing the final decision. Sir William Johnson, England's colonial Indian agent, reported the verdict of the Wappinger land case to the Earl of Shellbourne on March 5, 1767; the case, he said, "was vexatious" and "the Indians had no claim."

The governor ordered the Wappingers to move off the land, and the Philipses demanded that all former Wappinger tenants sign new one-year leases for increased sums, to be paid in cash—as opposed to the more tenant-friendly terms of a previous generation that accepted payment in the form of winter wheat, which had long been the tradition at Philipsburg Manor. Though Nimham would bring appeals until the Revolution, the episode had reached some closure, at least for Mary and her fellow heirs. In the course of the fight they had found that they could afford every weapon they needed to defeat land claims that challenged their supremacy. By the end of 1767, they could all move on, back to the more civilized activities that occupied the center of their lives.

*I*n 1750, when Mary was just out of her teens, painter John Wollaston had captured the self-satisfaction of her girlish face in a portrait. The portrait's almond eyes might have seemed generic, like all Wollaston eyes, as though she belonged to the Wollaston rather than the Philipse family, but Mary liked the painting well enough to render a copy of it herself (duplicating familiar or admired images was a frequent hobby among genteel colonial women). In a dainty mobcap, with a delicate embroidered rose at her bosom and a black velvet ribbon encircling her throat, Mary looked as if she were waiting for some suitor to come pluck her.

Years later another family portrait captured a more mature Mary. In the 1760s the popularity of Boston portraitist John Singleton Copley exploded among the cognoscenti of the colonies. Though still in his mid-thirties, the painter had already been elected a fellow of the Society of Artists of Great Britain. His fame had been launched by the submission of one of his canvases, portraying a mischievous boy and his pet flying squirrel, to the influential English painter Benjamin West. (The painting's frame of American pine was the only hint of its origin, and Copley did not come forward until it had been exhibited anonymously to acclaim.) Copley seemed to be painting all the best people, and his keen interest in fashion as well as art meant his subjects invariably were attired in the most up-to-date of costumes. When a woman sat for Copley, she knew she would be captured for posterity exactly as she wished to appear.

The white satin gown Mary wore in her portrait of 1771 was utterly à la mode—white as a fashion hue had only recently made the leap from the garments of children to those of adults. Her costume was adorned with an equally trendy embroidered belt with a Turkish accent. The picture's most striking feature, though, is the melancholy yearning in Mary's heavy-lidded brown eyes, an effect over which she had less control than her costume. Mary was a matron of forty-one, a calm, self-sufficient woman and something of a romantic. Looking around, playing her part in the world, she had begun to experience an uneasiness she had never counted on in her youth, when every comfort of Manhattan and Philipsburg was hers. But she remained a creature of fashion.

15

Fire in the Sky

*T*en years previous, Mary would not have believed the scene possible. Above the southern horizon, a red glow blossomed in the midnight sky.

It was September 21, 1776, and the city of New York was on fire.

From the Manor Hall's elevated site, you could clearly see billows of gray pump into the sky from the shadowy land below. When Mary ran down to the Strand along the Hudson she could see straight south along the open, starlit highway to where the island of Manhattan was shrouded in smoke. The sight was more terrifying than anything she had ever seen.

When Mary was a girl, her mother Joanna had dragged her by the hand to see the punishments of the black conspirators of 1741. (What a mother she had! Mary would never in all her modern, late-eighteenth-century, child-centered, affectionate, maternal perfection subject her own innocent babes to so grisly a sight.) All the wise men, her father included, said the threat of black men to the city was the worst imaginable.

Now, just six days after the Patriots fled New York and His Majesty's troops moved in to occupy the island, the city had experienced this calamity of unimaginable proportions. And though the identity of the person who struck the match never would be proven, it seemed obvious from the first terrifying flames that the culprit was a Whig.

The war had erupted the previous year. For more than a decade, while Mary built her house on the hill, colonists had protested British tax policies and boycotted British goods. Shipments from Britain dropped almost in half. Colonists convened the Stamp Act Congress to hatch polite appeals for relief from the Crown. The "Sons of Liberty" formed less-polite secret societies to protest the taxes. Tensions escalated when the Crown dispatched four thousand troops to America in 1768. Brawls ensued. In March 1770 news of a chilling episode raced through the colonies: When colonists pitched snowballs at British sentries guarding the customs house in Boston, the sentries fired into the crowd of several hundred, hitting eleven people and killing five. Afterward England lifted all the new duties except those on tea, an effort to soothe American tempers while sending a clear message that Parliamentary supremacy still reigned.

Tea would remain a lightning rod of controversy, a fact that held a particular resonance for a woman of the upper echelon who was more than a little wed to the rituals of the tea table. Female mobs styling themselves the Daughters of Liberty, sisters to the Sons, hurled rocks through tea shop windows. Rejecting the beverage became a fad after fifty men dressed as Mohawks dumped chests containing ninety thousand pounds of tea into Boston Harbor in 1773. But giving up tea was a struggle, especially in Anglophilic New York. Even in Boston, while the rebels pulled tea boxes out of the hold, townspeople crept on deck to scoop up fallen tea leaves. An eyewitness described a Captain O'Connor, who "when he supposed he was not noticed, filled his pockets, and also the lining of his coat." Once detected, the thief made his escape, but he had to "run a gauntlet through the crowd upon the wharf" as he passed, with each person "giving him a kick or a stroke."

Mary could understand the argument that drove this crusade—no taxation without representation—but the protest seemed to go too far. All this over imported tea! Female patriots had taken to brewing all sorts of local potions and giving them clever names: Liberty Tea was made from loosestrife, Hyperion from the dried leaves of the raspberry bush. Some women concocted tea from parts of the linden tree. The supply of sugar and coffee also dwindled because of their boycotts. Even Mary had to say that forgoing these luxuries made sense if it helped persuade the Crown not to tax excessively. But was it necessary to include salt in the boycotts? People said it was possible to preserve a ham with a lye extracted from walnut ashes, and she hoped she would not have to try it.

Some women vociferously declined to wear satins and silks because they too came in off merchant ships, subject to taxation. They wore muslin instead. They claimed to be happy wearing plain clothes, reserving their finest sewing efforts and best linen for soldiers' uniforms. One leader of the Daughters of Liberty in Massachusetts wrote of this new clothing style, "I hope there are none of us but would sooner wrap ourselves in goatskin than buy English goods of a people who have insulted us in such a scandalous way."

But Mary, who came to her money by way of old-fashioned merchant pragmatism, could also see the opposing view: Government taxes were the price of being responsible citizens. Many women who felt as she did made a point of wearing fancy dresses. They continued to patronize tea shops, hold on to their contraband sugar, and smuggle in luxuries out of pure defiance in the face of what they saw as silliness, or sedition.

Women threatened women with violence. When a Massachusetts woman bore a son in 1775 and publicly expressed her intent to name him in honor of British commander Thomas Gage, a crowd of female patriots descended upon her house, tar and feathers in hand.

Mary had named her own baby Henry Gage. General Gage and his American-born wife, Margaret Kemble—an heiress descended from Van Cortlandts—were particular friends of the Morrises, as were Lord and Lady Gage, the general's brother and sister-in-law. The fact that they were social peers made the friendship especially precious during the fractious mid-1770s.

All of the colonists knew about the violent acts over imports and tussles between the Sons of Liberty and soldiers. Yet the whole thing seemed somewhat extraneous to New York women who were equally loyal to His Majesty and to fine imported goods. Margaret and her contemporaries assumed these unpleasant trends would blow over and that the colonies would remain united under the Crown. Who could forget the close of the French and Indian War only a few years past? The colonists had come together to rejoice in the English victory and the new freedom from worries about foreign incursion. Mary had raised toasts, everyone toasted together, praising the king and William Pitt, the architect of the war victory. For that moment, the people of America were one with England. And they were unified among themselves.

Almost immediately, it seemed, the situation had escalated from dissatisfaction over taxes to the strife of battle. In 1774 Parliament laid down a series of measures that came to be known as the Coercive Acts. In an effort to regain control, the British closed the port of Boston entirely, made it impossible to try Britons for capital crimes in America, and basically annulled self-governance of the colonies. The colonists called a Continental Congress that same year with the purpose of rejecting royal authority. On April 23, 1775,

King George III declared that the "colonies are in open and avowed rebellion. The die is now cast." The battles of Lexington and Concord followed and the colonies formed a new patriotic army with a proven hero for a commander, none other than Mary's old swain George Washington.

Meanwhile, Mary's husband, Roger, took himself off to safety. On May 4, 1775, in the convulsive aftermath of Lexington and Concord, Colonel Morris rushed to get a spot on a mail packet for England, leaving Mary in charge of the children, the city home, the Highland acreage, and their brilliant aerie atop Break-Neck Hill. He explained in a letter that he did not want the Sons of Liberty to torch the family's property because of his presence in New York. (Though Roger did not declare himself a Tory, as a member of the governor's council his royalist convictions were assumed.)

Roger had siblings in London and knew he could count on a soft landing there. He would reunite, as well, with the eldest Morris daughter, Nancy, now almost fifteen, who had been shipped off to an English finishing school in June 1773. The girl had for two years now been cared for by the Gage household, which had a daughter the same age as Nancy and divided its time between an exceedingly comfortable London town house and an estate at Firle, in Sussex. Fortunate Roger would now get to see what Mary would necessarily miss: their daughter's blossoming command of the crucial elements of refined comportment, as schoolmistress Mrs. Cockerell brought her along in dancing, French, drawing, and geometry.

By early 1776 the rebels ousted all of the royal governors of the colonies and installed republican governments. In July 1776 the Continental Congress met again and adopted the Declaration of Independence.

Amid widespread excitement over the new republic, some colonists had different perspectives. In English politics, a Tory supported the king while a Whig was an opposition member of

Parliament. The same nomenclature clung to the warring parties in America. But, at least at the beginning of the conflict, allegiances were not quite so polarized, so simplistic. Colonists held a range of nuanced affiliations, especially in the heavily British, affluent, Anglican province of New York. The colony's population of roughly 168,000 residents broke down into a small number, probably 15 percent, who were hard-core Tories; 60 percent who were Patriots; and a balance whose stance was perhaps more practical than ideological. Some of the city's close-lying counties, Queens, Kings, and Richmond, were known to be Tory bastions. In Westchester the ratio of Patriot to Loyalist has been estimated as fifty-fifty. The Tories transcended racial, ethnic, religious, social, and economic bounds and came from all walks of society, as did the Patriots. Often households broke apart into Patriot and Loyalist camps.

The Philipses maintained at least a semblance of neutrality until war actually broke out. Although certainly Philipsburg and the surrounding terrain had its share of wealthy Tories in country houses, local loyalists were not exclusively rich merchants, landowners, and politicians. On the Philipse lands, some tenants were simply accustomed to English ways, to English rule, and felt it was safer to go with the familiar. Some feared the anarchy of mob rule. Others voted with their stomachs. A number of farmers felt a stubborn loyalty to the Philipse family, which had always provided for them.

It was not a simple distinction. Some Tories loathed imperial levies, while some Whigs had to be persuaded to split with England. Even as war loomed, some delegates to the Continental Congress pressed for a reconciliation with the Crown. There must have been a distinct anxiety in finding yourself between the two camps intellectually, especially when you were also torn between the opposing sides in a physical sense. All parties could agree that Manhattan was both a prize and a strategic necessity for both sides of the conflict.

From 1774 to the end of the summer of 1776, the Patriots had held sway in New York City. Revolutionary fervor produced patriotic rallies and raids of the city hall arsenal for guns and ammunition for the American militia. A General Association and a Provincial Congress scheduled their first meetings (as evidence of the blurry distinctions, some prominent Tories still took part). In April 1775, the Patriots shut down the port of New York until further notice. (The ship on which Roger escaped carried mail and so was exempted.) By the middle of May, Mary had ensured her offsprings' safety. Amherst, at twelve considered nearly grown, would remain safely at his boarding school. The younger children— Maria, now ten, and Henry Gage, five—she sent to the Highland Patent, where their aunt Susannah awaited them. It took until June for the last British regiment in Fort George to withdraw to a warship in the harbor, but not before the rowdy Liberty Boys stripped the troops of muskets and ammunition. During the American occupation, lukewarm Patriots fled the city along with established Tories, unwilling to be a part of the violence to come. Uncomfortable in town, Mary decamped for Yonkers, where Philipse Manor Hall remained a sanctuary from the heat and the threat of violence.

Meanwhile, the thousand-troop Continental Army transformed the city. The beautiful shade trees were soon leveled and American defense works clogged the streets. Soldiers squatted in the town houses of the elite. Valuables were stolen or vandalized. The Provincial Congress even asked homeowners to remove lead from windows, sashes, and drapery weights so that the army could recycle it to make ammunition.

In the fall of 1775 Mary heard that in her absence British guns had fired upon the city and that a shot had pierced the roof of her house on Stone Street. She stayed the winter with her brother and sister-in-law in Yonkers, keeping up her spirits by commissioning ornamental shrubbery for the grand boulevard that approached porticoed Mount Morris. The gardener there would prove thankfully loyal to Mary and Roger, staying on through the conflict to groom the gardens.

At the same time Mary staked out a position of caution, moving most of the furniture out of the Stone Street house and exercising a new frugality by selling two of her domestic slaves, Jack and Laba. She took special care to preserve the portrait her husband referred to as "Copley's Performance." When she came to Stone Street to supervise the loading of the carriages that would caravan the family's belongings to Mount Morris, lower Manhattan was changed, emptied and quiet, awaiting the worst. Mary then reclaimed her youngest children—Henry, spoiled by his aunt's attentions, and Maria, who practically glued herself to her mother and had grown distressingly thin. Amherst returned from school, an adolescent boy still but nearly ready to run off to whatever fighting unit would have him.

As 1776 began, the atmosphere of dread in the colony intensified. Patriot soldiers swept through the streets of the city in the winter of 1776, chasing down suspected Loyalists for questioning. By the end of spring, everyone expected an imminent British invasion. England viewed New York City as a key military prize, with its fantastic harbor and access to the two river highways flanking the island of Manhattan. His Majesty's troops had been busy in Massachusetts in preceding months. Patriot intelligence predicted the next engagement would bring the English south, to New York Harbor.

Mary would not be there. By summer she had fled, first for Mount Morris and then for Philipse Manor Hall. Even from that distance, however, she would have heard about the angry mob that pulled down the equestrian statue of George III in the Bowling Green, the park her father helped found decades earlier, just a block from the brick house she and Roger built together. Cast in lead and then gilded, the four-thousand-pound statue had been commissioned by New York City and created by one of London's most celebrated sculptors. Not long after its unveiling in 1770, political dissenters began to grumble about the imperial glorification in their midst—so much so that in 1773 an antigraffiti, antidesecration law went on the books to quell the vandalism that desecrated the royal likeness. On July 9, 1776, after the Declaration of Independence was read on the Commons in front of City Hall, a pack of Patriot soldiers and civilians had marched to the Bowling Green, dragged down the statue with ropes, and proceeded to dismember it. They cut off the king's head, mutilated his nose, ripped the laurels from his brow, and dragged his gilt torso through the streets. The statue, though, was not a total loss. The American army melted down its metal remains, and an estimated 42,088 bullets were ultimately cast from the bodies of King George and his steed.

Two months later, after the British came in, the unthinkable occurred: A devastating blaze ripped through the nearly deserted Manhattan. Accounts of the fire's origin differed, but some witnesses said the first flames emanated from Fighting Cocks tavern in Whitehall Slip, an establishment generally patronized by English sailors. A stiff wind from the south-southeast whipped the fire toward the densely packed buildings of the central city and the structures' cedar shingles made perfect kindling for the blaze. There were no bells to sound an alarm—George Washington had ordered them carried out of the city when his army vacated, their metal to be melted down for bullets like the shell of King George. Firefighters were scarce. As it happened, many were Patriots, who had already fled the island in anticipation of the British arrival. After twelve hours the raging blaze had laid waste to a full mile of Manhattan's west side, from the island's southern tip past Wall Street, and from Broadway to the Hudson River.

Mary did not know as she watched the smoke rise on the night of September 21, 1776, but the fire had ripped through the neighborhood where her family had carved out its Manhattan existence for a century. It consumed nearly 1,000 buildings, including 492 houses, landmark churches, and educational institutions—about one-quarter of the built city. Gone were pleasure gardens, taverns, waxworks, theaters, carriage makers, bake shops. Even the landmark spire of Trinity Church, now eighty years old, had toppled to the ground. "The Steeple which was 140 Feet high, the upper Part of Wood, and placed on an elevated Situation, resembled a vast Pyramid of Fire," one witness reported, "and exhibited a most grand and aweful Spectacle. Several Women and Children perished in the Fire; their Shrieks, joined to the roaring of the Flames, the Crash of falling Houses and the wide spread ruin which everywhere appeared, formed a Scene of Horror great beyond Description and which was still heightened by the Darkness of the Night." Mobs of English sailors and soldiers spread through Manhattan, looking to hang, shoot, and bayonet Patriot arsonists. They strung up men found with bundles of rosin-dipped matches or suspected of cutting the handles of leather water buckets. Petty crime followed, such as the theft of books from the city hall library.

Before the war the city had a population of about twenty-five thousand. So many people had fled New York—first Loyalists when the Americans held the city, then Americans when the British invasion loomed—that only five thousand souls inhabited Manhattan when the fire broke out. These people simply had nowhere else to go and no funds to leave town anyway. The fire killed women and children left at home when their menfolk went off to war. Many who survived lost their homes. A fetid tent city called Canvas Town sprung up west of Broad Street. Tory refugees—including those who had begun to return to the city after the British seized it, since it was now safer than the Patriot-leaning countryside—congregated there in huts fashioned by covering still-standing chimneys and brick walls with whatever old ship sails they could beg or buy. Wartime facilities were scarce. King's College became a hospital.

The English incarcerated prisoners of war in the hulks of decommissioned warships and in abandoned sugarhouses in Brooklyn's Wallabout Bay. There an estimated twelve thousand American soldiers perished over the course of the war in a welter of filth, contamination, starvation, and disease—far more than those who lost their lives in combat.

Mary was a prisoner of a different sort. Like many women whose male family members had marched off to fight, she remained at home to witness the war escalating outside her front door. In this case, her home was her childhood home. In September 1776, as the city went to ashes, she found herself stranded with her children in the midnight blackness of the Manor Hall, conserving candles. Roger Morris was still holed up in England. He had not overestimated the threat to colonists with royal connections.

As far back as 1766, the Sons of Liberty had forced their way into the opening performance at a new Manhattan theater. Shouting "Liberty!" they dragged members of the audience into the street, ripped the wigs off theatergoers' heads, and trashed the building, making a bonfire of its wood planks. The Tories dubbed them the Cudgel Boys. A more common tactic involved the contents of backyard privies, which the Cudgel Boys used to bespoil the exteriors of shops that sold English goods. Other times the Cudgel Boys tarred and feathered selected Loyalists. During a hazing, they heated pine tar to bubbling in front of the bound wretch's eyes. After stripping him to the waist, his tormentors poured the compound over the victim's chest, back, and shoulders with a long-handled ladle, sometimes soaking the head. Feathers were a grace note, comical to watch as the victim got rode around town on a rail.

The Sons did not fool around. Roger Morris considered himself a likely target and did not want his wife or children to lose him to the mob. If any less-noble explanation for fleeing surfaced in his agitated person, Roger did not express the sentiment to the beloved wife he left behind. He sent Mary rhapsodically homesick letters in his stead (preserved because he copied each epistolary gem before sending it on the packet to America). He would be gone for two years and seven months.

"Would to God, we were once more together; you cannot conceive, my dear Mrs. Morris, how much I think of you," he wrote. "Your repeated marks of tender Love & Esteem, so daily occur to my mind, that I am totally unhinged. Only imagine that I, who you very well know never thought myself so happy any where as under my own Roof, have now no home, & am a wanderer from day to day, & were it not that I am thoroughly convinced that no misconduct of mine has brought me to this disagreeable, unhappy & for the present unavoidable situation, I should be most wretched indeed."

Actually he did not wander but had "taken chambers" in London, where he sought out news from America as soon as any likely ship pulled into port. Mary seems to have written seldomly, but her husband had plenty of time to draft communiqués. "Dearest Life," went a typical Morris effusion, "my chief wish is to spend the remainder of my days with you, whose Prudence is my great comfort, and whose Kindness in sharing with patience and resignation those misfortunes we have not brought upon ourselves is never failing." Epistolary sweet nothings were not unheard of in this pre-Revolutionary era of increasingly companionate marriages, but Roger's tear-stained letters seem especially soggy.

As he wrote his beloved wife and waited out his self-imposed exile, Mary's husband passed his time paying social calls both in cosmopolitan London and to friends' fine country estates in Bristol, Teddington, and Oxford. Though his self-described routine was to "come home to my lodgings & sit moping by myself," he also managed to take the waters at Harrogate, a resort he deemed the most cost-effective in England for treating "scorbutic complaints," though it required a two-hundred-mile trek from London. He even accompanied a friend on a pleasure excursion to Gottingen, Germany. When he stayed in town, Roger kept a stern eye on the development of his already "too womanish" daughter, whom he wanted above all to learn how to "come into a room well." Nancy, he informed Mary, "is certainly my dear a fine girl (but I don't think her handsome) in need of her Mother's attention and care." Yet it still seemed too dangerous, Roger asserted, for daughter or father to return to America at present. Mary would wait years to reunite with her firstborn.

Meanwhile the other colonel, Morris's brother-in-law Frederick III, was beset by problems he had never in his life thought to anticipate. In March 1775 he had arrived at a White Plains gathering of freeholders intent on selecting delegates for the upcoming Second Continental Congress to let them know that he strenuously opposed moving forward with the meeting of that body. He had also signed his name to an advertisement that appeared in the *New-York Gazetteer* the following month professing support for the "King and Constitution." But those were the most extreme activities he had involved himself in as a loyal subject. His style of neutrality was expressed in the fact that he had been as shocked and saddened as any Patriot when he heard of the bloodshed at Lexington and Concord.

Then, seemingly oblivious to the potential consequences, Frederick all but guaranteed his arrest by calling together a crew of tenants on his comfortable Philipsburg home turf in the spring of 1776 to rally their support for the Crown. A new Committee on Conspiracies put Frederick on its "List of Suspected Persons," twenty-one Westchester Tories who must immediately be summoned and imprisoned. In the summer of 1776 he was escorted to Middle-town, Connecticut, and held as prisoner under house arrest.

His letters home to Elizabeth suggest that the conditions were less comfortable than Frederick would have wished. (He had already written to the Congress hoping to recuse himself, complaining of blindness in his left eye, with "the other so much inflamed as to make me very cautious how I expose it, for fear of a total loss of sight.") Unlike less-privileged prisoners, Frederick could make special requests of Elizabeth, to be facilitated by the corps of slaves at their disposal: "A few good Lemons would be verry acceptable by [enslaved] Diamond," he wrote in one letter.

While interrogators grilled Frederick in Connecticut and a special subcommittee of the Convention of Representatives of the State of New York branded him the worst kind of traitor, Elizabeth received her husband's letters at the Manor Hall, where she was left to care for her nine children in the middle of a war zone. She had for adult companionship only her sister-in-law Mary—who materialized regularly whenever the going got rough in Manhattan—and the servants who remained at the Manor Hall. Mary's brother Philip had died of natural causes in 1768; Frederick III was in prison. Roger was still in self-imposed exile in England and did not know that his wife and children had fled Manhattan. In the early fall of 1776 Mary and Elizabeth and their children clung to a reed of hope that all would be well.

In Mary's absence from her white castle, an old suitor moved in to fill the vacancy. General George Washington, no longer the young buck of her casual flirtations, made Mount Morris his headquarters after pulling his troops out of lower Manhattan on September 14. Mary's house, with its comfortable dimensions and long-reaching vistas, was the perfect site from which to wage a military campaign. His presence there was an imposing one, and she could imagine his six-foot frame striding across her octagonal drawing room. He was known to crack Brazil nuts between his jaws, a rare affectation in a man known for his gravity and physical calm. (It was also the habit that would cost him all his teeth but one by the time he became president, necessitating multiple dentures that utilized carved wood, cow's teeth, and elephant ivory as well as human teeth. The most ingenious was probably the full set of choppers carved from a hippo's tusk, with a slot to accommodate that single tooth; the pain of wearing this apparatus drove Washington into the arms of laudanum.)

While ensconced at Mount Morris, the general paced the grounds behind the house, the unembellished, clapboard side, not meant for fancy guests, or ascended the balcony, spyglass to somber eye, taking in the stupendous panorama that encompassed so many landmarks of New York. The general was forty-four. Like Mary, he had a perfect view of the red glow that rose above Manhattan in the midnight hours of September 21, as he stared down toward the smoky horizon and surmised what had happened.

Correspondence and journal notes from that period reveal that Washington's superiors explicitly prohibited the Patriot Army from using arson as a weapon on Manhattan. Nonetheless, some of Washington's men had been tempted to avenge their ouster from lower Manhattan. Their desire for retaliation was further fueled during the Battle of Brooklyn just days earlier, in which a British force easily took down the inexperienced American soldiers. The ninety-five hundred troops left under Washington's command had made it back across the East River from Long Island, a legendary retreat under cover of fog and night. But Manhattan offered no safe haven. British gunships hovered like horseflies on the corpus of the East River, as they had for weeks, keeping up a barrage of fire. Cannonballs tore holes in buildings; one plowed into the ground within six feet of Washington as he rode into the fort. The need to retreat was clear as soon as the battle was lost—to stay would be suicide.

The Morris estate's transformation to a bastion seemed almost natural. General Washington filled the house Mary built with his "military family," as he termed his closest staff. They had plenty of good furniture for their use, as the house was stuffed with the pieces Mary had removed from Stone Street. Mount Morris made a deluxe encampment; soldiers unfolded their rickety cots to sleep in the airy central hall upstairs, and paraded before the majestic front portico at dawn. The brigade-majors of Washington's thirteen brigades arrived during the course of the day to deliver their reports. Mary's formal parlor hosted trials by court-martial. The adjutant general operated out of the front of the house, while Washington and his aides took over chambers on the second floor, signing orders as the nighttime breezes flooded the cavernous rooms, with a typical dateline reading, "Head Quarters, Col. Roger Morris's House, ten Miles from New York." Ebenezer Hazard, the postmaster of New York, also kept his office on the grounds. The dining room retained its original function. As was the custom for Virginia gentlemen, General Washington dined every day at three, with six staff members, the brigadier, the officer of the day, and other distinguished guests. The basement kitchen with its ferocious oven was pressed to the limits of its ability.

*A*fter September 1776, Mary remained at the Manor Hall, though she managed one trip into Manhattan after the fire to assess the damage to her property. There Mary witnessed the worst, the specific devastation that until then she had only heard rumors of: The great fire had consumed the house she and Roger built on Stone Street. She had no more home in Manhattan, nor could she find a haven at Mount Morris. The estate had been appropriated by the Crown after the Battle of Fort Washington in November drove away the Patriots. At various times, Mary's seven-bedroom mansion hosted Sir William Howe, General Sir Henry Clinton, Oliver De Lancey, and Hessian General Baron Kniphausen. Her petitions to reoccupy her property received no encouragement from military officials. The house's siting was simply too ideal for His Majesty to relinquish (though Mary did receive one hundred guineas for the inconvenience of being evicted). As 1776 turned to 1777, she and the children were forced to make do in Yonkers. Margaret's storehouse, and Joanna's party house, had turned into a refugee camp.

Mary's privation was not total. Some things did not change, despite the war outside. The family continued to experience uncommon privilege. They still had fine clothes, beautiful furnishings, and a staff of servants and slaves. They raised farm animals and vegetable gardens, so meals included the basic menu items they enjoyed. (Their mills still produced all the flour they could want, for all the bread they could eat.) The central change was a loss of liberty—it was no longer safe on their own property to walk or ride, even to pick apples in the orchard below the house; all of Westchester had become a battlefield.

Termed the "neutral ground" after the Battle of White Plains in late October, 1776, the lands to the east of the Hudson actually represented a free-fire zone. The guerrilla groups known as the Cowboys (Loyalist) and Skinners (Patriot) rampaged at will, along with the British, American, and French troops, and assorted privateers, common thieves, and criminals. In a letter of November 7, 1776, Washington complained of the British army's actions as the force moved through Westchester, writing, "They have treated all here without discrimination. The distinction of Whig and Tory has been lost here in one general scene of ravage and desolation." He was more specific in a letter to General Nathanael Greene: "Many helpless women had even the shifts taken from their backs by the soldiers' wives after the great plunderers had done." Clearly, female camp followers played a part in the plunder, if only as beneficiaries.

Colonists especially feared the Hessians, German mercenary soldiers in the service of the British. The residents of Philipsburg viewed them as less civilized than other soldiers, saber-wielding bullies, prone to pillage. Rape threatened women in all colonies throughout the war, and it's easy to see a tinge of menace in the observation of one Hessian officer about the American women he saw. "They have small and pretty feet, good hands and arms, a very white skin, and a healthy color in the face, which requires no further embellishment. They have also exceedingly white teeth, pretty lips, and sparkling, laughing eyes."

When soldiers asked residents to open their homes to quarter troops, a frequent occurrence, there was no way to decline. Westchester residents attested that officers of both armies stole furniture, clothes, cattle, and horses. Soldiers removed fence rails for bonfires, letting loose cattle. During the Battle of White Plains, one family reportedly herded its milk cow down to their basement to keep it from being a soldiers' barbecue. After that engagement, the British controlled the southern portion of the manor, and Lieutenant-Colonel Stephen Kemble recorded his displeasure about his own men in his diary: "The Country all this time unmercifully pillaged by our Troops, Hessians in particular; no wonder the Country People refuse to join us."

On November 17, 1777, two Loyalist captains led a raid on the homes of the patriot Van Tassel family, a fixture of Philipsburg. In what was later termed a retaliation for the torching of New York, the men burned the houses, "stripping the women and children of the necessary apparel to cover them from the severity of a cold night," and took the men as prisoners, increasing the humiliation by tying their hands to their horses' tails.

Elizabeth and Mary seem to have taken an ecumenical approach to the conflict. In a letter to the American Colonel Webb, a military postmaster responsible for the transmission of mail to and from

Frederick's Middletown prison, Elizabeth expressed her profound gratitude about his promptness in forwarding the correspondence between her and her husband. Next she "begs her compliments to General Washington." Her tone was a relic of the ties that had always bound colonial men and women of a certain class. She also wrote Washington directly, complaining that the soldiers were driving off her cattle.

Washington's October 1776 response had gentlemanly diction: "The Misfortunes of War, and the unhappy circumstances frequently attendant thereon to Individuals, are more to be lamented than avoided; but it is the duty of every one, to alleviate these as much as possible. Far be it from me then, to add to the distresses of a Lady, who, if I am sensible, must already have suffered much uneasiness, if not inconvenience, on account of Colonel Philips's absence." Sweet nothings from the man who ordered that absence! But the general added a postscript: "I beg the favour of having my compliments presented to Mrs. Morris." After living in her house, sleeping on her linens, and imbibing wine from her cellar, it would seem Washington would have more to convey to his one-time crush than that. But such were the compromises of wartime.

In November Elizabeth tried again to get some satisfaction about the cattle, which were now being impounded to Peekskill, a village on the Hudson about twenty-one miles north of the Manor Hall. She complained to General Charles Lee, who responded first with the patronizing assurance that her herd had to be impounded by the Patriots lest "Hessian Marauders" got to it first. When she persisted, he lectured her that "Neutrals in civil country can by no means be suffered," and then threatened, "those that do not take a positive active part with their Country [shall be treated] as enemies—their whole stock shall be seized & their houses burnt."

It was certainly becoming difficult for Elizabeth and Mary to project a neutral stance. It was natural for some Patriots to suspect

Mary's position, especially because her council-serving husband had been born in England, fought for England in the last war, and was in England now. Yet even with Frederick under guard in Connecticut for his Loyalist convictions, Mary and Elizabeth behaved politely to any officer in the vicinity of the Manor Hall. They would offer hot chocolate to the American officers—and tea from the locked tea box when the Tories dropped by. But there would be no bland muslin gowns, that was for sure. Mary might feel a little funny, dressing her best with New York burned to a crisp, but the English soldiers made it seem all right. And really, looking well was all she had at the moment.

Roger had written his wife in June that they could always, if desperate, move to "our log house, [which] will then be a Palace for us, and I hope to live as comfortably . . . as ever we did." But for now that would not be necessary. Frederick Philipse III came home from Connecticut on December 23, 1776, after a six-month imprisonment, just in time for Christmas. His release on parole was governed by three terms: He could not give intelligence to the enemy, take up arms against the Americans, or do anything inimical to the Patriot cause. It seemed unrealistic, the idea that the obese and nearly blind Frederick would stir up trouble in any way. Surely he would go back to his flower beds—to the extent possible in a landscape crowded with soldiers—rather than engage in any kind of potentially damaging activity.

Within six months of returning home, however, Frederick III spied a column of Connecticut soldiers moving south past the Manor Hall. He rushed off a note to the British at Kingsbridge saying the soldiers were likely on their way to ambush the British encampment at Morrisania. The communication was intercepted, the violation of his parole self-evident. While some contemporaries blamed his act of sedition on the influence of his wife, "an English woman of strong Royalist sentiments," Frederick might have been motivated by disgruntlement over spending an uncomfortable six months in jail. In any case, the die was cast. Yonkers did not seem the wisest choice of residence just now for the beleaguered family, especially with nearby, familiar Manhattan under sympathetic British control. Accordingly, the entire crew—Mary, Frederick, Elizabeth, and all their children—moved quietly down to New York on the Hudson, under cover of night, away from the hell of their own homeland. Not much later an American commander offered his perspective on the threat to certain elite properties as war engulfed southern Westchester. "You can not be insensible," he said, "tis every Day in my Power to destroy the Buildings belonging to Col. Philips."

With their Hudson River exodus, Mary and her family joined the flood of Tory refugees who returned to British-occupied New York in the months after the 1776 fire. While her mansion stood in ruins, she could feel fortunate that all family members were safe, and she had the means to let a house on Queen Street. She could even furnish the house with the things she had been prescient enough to remove the year before.

Once again, Mary put in motion her caravan of household goods. This time, the fancier furnishings from Mount Morris and the more rustic appointments of the Highland log house came in tandem to lower Manhattan, to make a jumble of the cramped dwelling the Crown had seen fit to assign her. The venerable family mansion still stood nearby, so Frederick and Elizabeth enjoyed the security and comfort to which they were accustomed amid the charred ruins of downtown. Frederick received a royal pension because he had been obliged to leave his estate and property on account of his attachment to His Majesty's government, a tidy annual sum in the neighborhood of two hundred pounds. Mary also saw her own income much reduced, with no Patent rent rolls or income-earning spouse at hand.

The British would hold the island for seven years, until the war ended. The British army apparently lived well during that time—officers were quartered at the lovely Ranelagh Gardens, and the Crown generously shipped over three thousand prostitutes for the troops. By February 1777 the city's population had swelled back to eleven thousand. The unluckier refugees had to make due with accommodations in the homes of vanquished middling Patriots (better certainly than Canvas Town), although their borrowed quarters on Manhattan were undoubtedly more elegant than the ones some left behind.

Roger Morris finally returned to New York in December 1777, after the English had occupied the city for more than a year. His eagerness to come home had been offset by a fear of abduction by hostile forces en route, but he made it in one piece. So did the new material goods he shipped. In the gathering storm of war, Morris had managed to commandeer dessert china, ribbons and handkerchiefs, a silk nightgown and a "handsome & fashionable" petticoat for Mary, along with a dozen hams, two dozen tongues, rose water, raisins, and a chest of tea. Seemingly clueless about the grave state of affairs in the countryside surrounding New York, Morris even sent two hundred sheets of window glass across the ocean to fix any panes that might be damaged at Mount Morris.

Now Roger rejoined Mary and their children in Manhattan, where he importuned the military government for a position as Inspector of the Claims of Refugees, a post held by his brother-in-law before him. He did not land the job until 1779, and his conduct while employed seems to have been less than exemplary. In January 1782, 361 refugees complained to the Crown about his "total ignorance of the character of the refugees," and "austere manners." Despite their plea for the government to remove him, the colonel stayed on until the end of hostilities in 1783.

The British in New York were joined by an unexpected addition to the populace: the freed slaves of Patriots. In the chaos of war, with all the passionate rhetoric on the part of Patriots about the rights of man, thousands of African Americans bet on their future by siding with the English. They had a very specific reason: In Virginia in 1775, the Royal Governor John Murray, the Earl of Dunmore, had issued a proclamation declaring that England would make all slaves and indentured servants of Patriots "free, that are able and willing to bear arms" for His Majesty's troops. The gesture stemmed less from ideological conviction than the Crown's war needs; in the phrasing of one British officer, black employment would "save the troops much Toyl and Fatigue." Nonetheless, the move represented the first whisper of black emancipation on American soil. And the Patriots made no such offer.

Thousands of enslaved blacks from throughout the provinces made their way to Manhattan, where they went to work for the British army, which proved less inclined to assign them arms than to have them dig ditches or serve as cart drivers. Mary and Roger had at least one slave, Martha, appropriated by the Americans when Mary vacated Mount Morris and Washington moved in; household receipts during the American occupation show "Marther the Negro wench" was paid for washing the general's linen. In lower Manhattan, the streets were filled with African Americans whose circumstances ran the gamut. There were newly free, army-employed blacks; individuals who were still enslaved by resident Loyalists (the Crown went out of its way not to deprive its supporters of their chattels); runaways who had breached enemy lines to escape their Patriot or Loyalist owners outside Manhattan; free blacks who had already resided in the city; and blacks left without masters when their Patriot owners fled. All the while, Manhattan's slave market continued to thrive.

Nor did the war halt the New York social scene, its denizens seemingly oblivious to the carnage in the fields and forests just north of Manhattan. A family such as Mary's existed in the city as if it were a kind of political air pocket, a safe spot of English protection, knowing that if they set foot elsewhere in the colonies they would certainly suffocate. The Loyalists waited the war out on Manhattan believing it was only a matter of time before the most magnificent imperial army on earth would triumph over the ragtag guerillas and they could take back their lands and their lives. In the meantime, affluent residents enjoyed the usual luxuries. There were handsomely uniformed soldiers—Frederick purchased five commissions in the British army for his and Elizabeth's five sons—and a lot of cash pumping into the city. The balls and parties, according to contemporary observers, glittered even more brightly than they had at any other time in Manhattan's existence.

One of Elizabeth and Frederick's daughters, the lively Maria Eliza, even got married in British-held New York in 1779 (to Lionel Smythe, seventh Viscount Strangford). Aristocratic society regarded the union as a brave vote of confidence in the future. Attire was not a problem: Merchant ships could still enter New York Harbor with supplies from Europe—most importantly, with new gowns from France. Dresses were concocted of sumptuous materials, like the head-turning 1780 number that has come down to us in the collection of a contemporary textile collector. Sewn of a stuff called "silver tissue," a stiff silk with gilt thread woven in minuscule dots from selvage to selvage, the gown was trimmed with silver lace, sequins, gimp, and tassels. Silver tissue was fit for a queen; Princess Anne had married William IV, Prince of Wales, in a dress with a six-yard-train of the fabric in 1734, while in 1761 a diamond-garlanded Princess Charlotte shone in a silver gown that matched the silver-trimmed suit of her groom, George III.

Royal events still provided excuses for extravaganzas in Manhattan. Seventeen-year-old Prince William Henry, the third son of George III, sailed into New York Harbor in September 1781 to an explosive royal salute and the manic cheers of spectators, especially those of the young, elite, female persuasion. Manhattan residents saw his visit as a sign that the Crown took the war effort seriously. Their satisfaction was compounded by victories to the south, as England captured the American forces in Savannah and Charleston. For a year, the prince's rock-star presence inspired formal dinners, parties, and fireworks displays. His entertainments included a trip to Beverly, at Garrison, which had not yet been impounded by the Americans (whether by fluke or because Virginians Robinson and Washington went back a ways can only be surmised).

The birthday of the prince's mother, Queen Charlotte, in January 1780 had already occasioned some wretched excess. After gun salutes in the harbor throughout the afternoon, guests took carriages to a ball at Hick's Tavern hosted by the Baroness Riedesal and General Pattison. It began as a modest celebration of the royal anniversary—without the queen herself, for she was home in England. But then at midnight, after hours of dancing and drinking, the gentlemen and ladies repaired to a dining room decked out in parterres and arbors of natural and artificial flowers, china images, and other tokens. And the baroness, masquerading in elaborate costume as Queen Charlotte herself, took her place on a throne under a canopy to receive her subjects. The ball was said to cost more than two thousand guineas (equivalent to more than two hundred thousand dollars in 2006 currency) and the whole soirée struck the editor of the *New York Mercury* as not only "the most perfect hilarity," but "the most truly elegant ball and entertainment ever known on this side of the Atlantic."

North of the city, the war raged on. Patriot troops made themselves comfortable on the grounds of the Manor Hall and no doubt in its elegant chambers as well. A unit of Hessians called the Chausseurs advanced through lower Philipsburg in September

1778, spending a rainy night that a soldier named von Krafft described in his journal: "We were near what was called Phillip's House, we opened the pretty church there and quartered ourselves in it [. . .]—in all about 30 men. Finally a search was begun and a large potato-field was cleaned out and many other luxuries brought in. Fowls, pigs and beef slaughtered, although everything needed to be done secretly." Plundering had been strictly forbidden. All in all, he wrote, the Philipse grounds were "good night-quarters" for the mercenaries. A year and a half later, sixteen thousand British troops pitched their hut camp on the grounds of the Manor Hall. It was June 23, 1780, and they disembarked their warships at the mouth of the Nepperhan, which had served for so long as the ideal little harbor for sloop exports of grain.

Some men in the extended Philipse family did not delay declaring their loyalties in the war. In spring of 1777 Beverly Robinson left Susannah in the safety of British-held Manhattan and raised a company called the Loyal American Legion to fight for His Majesty. (He is recorded as basing his loyalty strictly on which political entity would allow him to retain the most property, a pragmatic admission in a conflict based so much on principle.) Four of his five sons fought under him, and the fifth served the British in another capacity. Not only did Colonel Robinson participate in some of the major victories of the New York theater, including Fort Montgomery and Fort Clinton in October 1777, he also performed as an agent for British intelligence. Tenants around the Robinsons' estate at Garrison remained staunch Loyalists like their landlord, though most did not stick around for the Patriot victory in 1783, instead fleeing north with many other Tories to Nova Scotia.

Frederick III continued to provide for Elizabeth and his children, supplementing his government stipend by collecting his tenants' rent—a seemingly arduous task while he remained behind the lines in Manhattan. Still, records show that as late as 1778, seventy-three Philipsburg residents actually made their way south from Westchester, through the lines, to settle up with the colonel at his comfortable home in lower New York.

*I*t was while she resided in Loyalist-protected Manhattan that Mary received word that there would be no more manor.

On October 22, 1779, the state legislature at Kingston signed into law the New York Act of Attainder, or Confiscation Act, which singled out fifty-nine individuals as "adhering to the King with intent to subvert the government and liberties of this State and the said other United States, and to bring the same into subjection to the Crown of Great Britain." Treason, in a word. Lawmakers settled on a punishment that made particular sense in the landed realm of so many New York Loyalists. The fifty-nine Tories would be summarily divested of their property in its entirety. Why those fifty-nine? The selection has been debated over the years, but it is clear that the legislature dealt a disproportionate hand of woe to the extended Philipse family, which had six members on the list.

The state named some actively treasonous individuals—martial types such as Beverly Robinson, who led men into battle and conspired with spies on gunships. But longtime council members such as Roger Morris and Frederick III, with his spontaneous tattling on the Connecticut regiment, seem to have been included for lesser reasons. His son and heir, British officer Frederick Philipse, Jr., also stood accused.

The legislature also targeted an exclusive group of women. Throughout the conflict female Tories had served as saboteurs, soldiers, spies, and even couriers for the Crown. The British Secret Service even sent one anonymous Loyalist to spy on Washington in Westchester; they were then officially considering kidnapping the general. She performed her assignment with gusto, according to one officer: "The woman is returned from Washington's Quarters. She saw him herself and says—that Washington sleeps in the back room—that there were two french sentries yesterday at his door and that his Guard consisted of French and rebels, which she judged to be about 30 or 40 men—that several Tents were near and about his House—that his Guard was much stronger by Night than by Day— that she saw no Horsemen there."

But the names of female spies and saboteurs were not put forward now. Instead, lawmakers targeted just three women, all from the finest families, with Mary topping the list. She was followed by her sister Susannah and Margaret Inglis, the wife of prominent Anglican clergyman Charles Inglis, who had become known for a clever attack he penned against philosopher Thomas Paine. Given her husband's energetic support of the Crown, Margaret Inglis's attainder was perhaps to be expected. But why were Mary and Susannah singled out? If treasonous perfidy meant marrying a man who professed allegiance to the king, then many more wives of Loyalists would stand accused. But the Philipse sisters possessed what the new government coveted—some of the most magnificent properties in all colonial New York.

As part of the 1779 Confiscation Act, the state planned to seize and auction off all Philipse "real and personal estates" in New York at the close of the war. This meant not only Philipsburg proper, but also the extensive Highland Patent acreage, all the New York residences, Mary's mansion on the hill, and any other random scraps of land the family had picked up along the way. But property would not be the only casualty of devotion to the Crown. The measure also declared that "each and every one of them who shall at any time hereafter be found in any part of the State shall be and are hereby adjudged and declared guilty of felony, and shall suffer Death as in cases of felony, without benefit of Clergy." In plainer language, any of the fifty-nine could be executed if discovered within state borders.

For the moment, the Philipse clan was safe in its Manhattan bubble. They saw no imminent reason to flee the colonies—but it was still necessary to weigh options. Some Loyalists had already left, but with so many valuables at stake Mary and her siblings would hang on until the bitter end—if you could call the New York syllabub-and-minuet circuit remotely bitter. It would be worse to leave the game too early—to ditch the lands, houses, and mills while there was still even a remote chance of getting them back with a British victory. The Philipses had built their fortune over a century's time, and they were not about to abandon it on the strength of a mere presumption that they might not be on the winning side. In 1779 it still looked as though the British might win this war. In any case, they knew that nothing would happen to Philipsburg Manor until the conflict concluded. They would ride it out.

Gradually, though, as the years went on, it dawned on the more realistic among the Philipse crew that their luck would not last. In 1781 the British saw the defeat of its last major army. The entry of France on the side of the Americans had proved decisive. It took two years more for the peace to be settled, and on September 30, 1783, a treaty signed in Paris confirmed the Patriot victory. In the immediate aftermath of the accord, Tories scrambled to find a way of staying—or a way of leaving. Neither alternative seemed optimal for the Philipse family. At the end of winter, when the ice melted and the first ships could leave the harbor, Loyalists began to embark for Nova Scotia, the Bahamas, and England; by August eighteen thousand had bid New York good-bye.

Relocating was more complicated for the Philipses. The source of their income was in limbo. Their confiscated property had not yet been authorized by the state legislature for sale; Philipsburg sat tantalizingly close to Manhattan, still intact yet completely inaccessible to family members. Rents throughout Philipsburg and the Highland Patent had represented a huge proportion of the income of every Philipse; without those rental properties they could not hope to survive in America as before. Frederick, Mary, and Susannah saw signs of hope in one of the peace treaty's articles, which stated that Congress would recommend that the states return material goods to their Tory residents. On the other hand, pamphlets and broadsides exhibited a public sentiment that was much less forgiving. One publication, by an author who identified himself only as "Cives," addressed Frederick III directly, expostulating that "the worthy inhabitants of Philipsburg who have at all times disputed that ground inch by inch, with the enemy, and purchased it with the price of their best blood, will never become your vassals again. They will not submit to become tenants *at will,* to you or your son, nor to any other *enormous landholder on such base terms.*"

Not all Tories would have to leave their homes in America, not even those suspected of treason. The new regime even invited some landowners to buy back their forfeited property. The New York State Assembly still included members of the best families, men whose private royalist leanings were clear but who had managed to avoid throwing in their hats with the losing side (hardly an option for blurt-prone Frederick). New York society welcomed back some Tories, although by law they could neither vote nor hold public office. Clemency, however, was a postwar gift to be bestowed upon some few, fortunate New Yorkers—those, unlike the Philipses, without large amounts of desirable property. As Beverly phrased it when he analyzed his prospects in 1782, his course would be determined not only by the fact that he and his family had been good Loyalists, but that they "had too good an estate ever to expect forgiveness."

There came a time when it was clear that forgiveness would not be extended to the Philipse family; that they could not stay. Then each household put up their goods for auction. Frederick, lord of the Manor of Philipsburg, got as much as he could for a japanned chest of drawers, among other items, while Beverly purveyed his personal oak writing desk to the highest bidder. On May 19, 1783, Roger and Mary took out an ad in the *New-York Gazette and the Weekly Mercury* that listed select items from all three of their residences. Shoppers visited the family's small Queen Street rental house to handle their goods: chairs and settees, bedsteads and looking glasses, desks and girandoles, a carpet, a bookcase, a clock, pictures, and even window curtains.

Without their land, a death sentence hanging over their heads, the family came to terms with its fate. Mary and Frederick III sailed to England along with their spouses and over a dozen children. They left on June 7, 1783, with the final departure of British troops—the harbor crowded with outward-bound warships, scarlet-uniformed soldiers jamming each deck—and there must have been a sense of anguish over having to evacuate this beloved place. An acquaintance would later recall meeting Frederick just before he left for England. "He was glad to see me," said General Pierre van Cortlandt, "but cried bitterly when he said, 'I must leave my country.'" Susannah, meanwhile, accompanied her husband to New Brunswick, Canada, an English colony that could accommodate loyal soldier-statesmen such as Beverly Robinson; they would eventually relocate to the English town of Thornbury, near Bath. Mary and Roger would settle in Saviour-gate, a village in Yorkshire, the Morris family seat. They would immediately press their claims with the Crown for reimbursement of their property losses.

Mary was still young enough at fifty-five to feel hope about this new life, even as she stood at the rail of her crowded ship of refugees and surveyed the shoreline of England, a country she had never before so much as visited. She was mature enough to trust that she could master this totally different world and make it hers, construct a life here the way she had built her white house on top of the hill. She would keep herself busy. Mary had her husband back, and her four children, now ranging in age from Nancy, twenty, down to Henry, thirteen; all needed her active care, even if that care now mainly meant ensuring that they married well. Amherst was already a man, having received his commission as a midshipman in the Royal Navy during the Revolution, in 1778. The girls she hoped would find officers of their own to wed, just as she had.

Mary had brought some of her beautiful possessions, cushioned in straw in the hold of the ship, that she had refused to sell. She and Roger had managed to pack out "Copley's Performance," which had been boxed away in one of the family homes for safekeeping since 1775. Crowded with refugees, the vessel could not accommodate every valuable Mary and Roger wanted to take, but they had been assured a stipend from the Crown that would keep them in a modicum of comfort.

Mary knew that the Highland Patent would never again shower profit on her family. Nor would she ever revisit the relatively brief pleasures of Mount Morris, the estate she had delighted in sculpting from a common farm atop a majestic promontory. As for Philipsburg, she could not help but be especially discouraged by developments there. It was not that Mary objected to the distribution of farmsteads to the residents of the manor; if tenants' outright ownership of land was indeed the American way of the future she would not begrudge them. But Philipsburg, its forests and meadows, high bluffs and wheat fields, the Nepperhan and the Pocantico, all seemed of a piece to her, an organism that would remain alive only if it were preserved intact. And the Manor Hall her great-grandmother built—a house for which the Forfeiture Commission would almost certainly never find inhabitants with a dedication to rival that of Mary and her family—the Manor Hall was Philipsburg's now-stilled brick and rubblestone heart.

Afterword

$\mathcal{M}$ary survived Roger Morris by thirty years, not succumbing until 1825, when she was ninety-five years old. Her sister and sister-in-law also outlived their husbands: Susannah passed away in 1822 at ninety-four, three decades after Beverly's death, and Elizabeth died in 1817 at eighty-four, thirty-two years after burying the blind and ailing Frederick. By the time of Mary's death, the family had become well rooted outside America, with its children married to Britons of aristocratic origin and their children raised as English citizens. Mary's youngest, Henry Gage, would become a vice admiral in the British navy; his older brother Amherst would rise to become a commander before his untimely death at age thirty-nine. While Nancy married in 1787, at twenty-six (like her mother coming close to spinsterhood before making the match), she bore no children, and Maria spent her seventy-one years mateless. Henry Gage supplied the genealogical juice for the group, producing a dozen children, including one Frederick, one Adolphus, one Joanna, and one Beverly.

A few family members ventured back to America over the years. But most continued to make their home where the fit was better for the family's Tory gentility. The Philipse bloodline survived in the United States mainly among branches that had traded in their allegiance to the Crown for loyalty to the new republic. The descendants of Eva Philipse and Jacobus van Cortlandt, for example, kept the family property in the Bronx. Mary's brother Philip Philipse, who did not live to see the war get under way, never had a chance to declare himself for the Crown, so his wife and heirs were permitted to retain his Highland Patent acreage and to deed it to future generations of American Philipses.

But these were exceptions. For the most part the Philipse family simply disappeared from the history of Dutch-founded New York, as though its century-long run in the region had never happened.

Some of the places they inhabited still exist, though.

Mary's hilltop castle survived the war intact. In 1790, less than ten years after her exile to England, General Washington returned to Mount Morris to host a commemorative dinner for a notable collection of guests. Today the house has been lavishly restored to its colonial appearance, making it easy for a visitor to visualize Washington regaling Thomas Jefferson and Alexander Hamilton in the grand octagonal withdrawing room. By that time Washington had married Martha Custis, another charming landowner's daughter with plenty of money. Soon the American government sold the mansion it had confiscated and Mount Morris became Calumet Hall, a popular tavern that served as the first stop for travelers leaving New York on Albany Post Road. Its owner, named Talmadge Hall, touted his establishment's features in an ad that appeared in the *New York Packet* in May 1785: "The Octagonal Room is very happily calculated for a turtle party, and [guests] shall have for desserts Peaches, Apricots, Pears, Gooseberries, Nectarines, Cherries, Currants and Strawberries in their Season." Eventually the property was bought by Eliza Jumel, a teenage prostitute from Rhode Island who became one of the richest women in America following the death of her wine-magnate husband. A passionate Francophile, she reposed on a bed that reputedly belonged to Napoléon himself. After marrying and divorcing then Vice President Aaron Burr, she lived at Mount Morris until her death in 1865. The mansion's sequence of moneyed owners ensured that the structure remained well preserved as the city grew up around it, and its extravagant profile still commands a 360-degree perspective, at least in winter. Mary's new, modern house is now Manhattan's oldest.

While modernization long ago replaced all the Philipse residences in Manhattan with high-rise buildings, the layout of the downtown streets they inhabited remains almost exactly as it was under Dutch rule. You can still stand on the northwest corner of Stone Street and Whitehall beside the very lot awarded Frederick Philipse for his service to the infant colony. If you close your eyes, the shadow cast by the structure across the street, the behemoth U.S. Customs House, could almost be that of the military fortress that Margaret, Catherine, Joanna, and Mary confronted each time they stepped out the front door of the family town house.

Another home still stands, beside a former cornfield in the Bronx on the property Frederick conveyed to Eva Philipse and Jacobus van Cortlandt in 1699. This is the stone mansion Eva's son Frederick van Cortlandt built in the 1740s as he competed with his cousin Frederick II at Philipse Manor Hall to establish which man could make a more impressive showcase for his fortune. The dwelling's drafty double-hung windows, thoroughly modern when they were first put in, overlook a last vestige of early New York:

a section of ungraded dirt path that was once a leg of the Albany Post Road. In the eighteenth century, Joanna Philipse would travel this rough trail through the forest to make her way north to Philipsburg—in a rocking carriage, attended by liveried slaves—after paying a call on her relatives.

After the Revolution, when it came time to supervise the dismemberment of Philipsburg Manor, the job was performed by another product of the Philipse and Van Cortlandt bloodlines—Philip van Cortlandt, the great-grandson of Stephanus and the great-grandnephew of Jacobus and Catherine. He served as one of two Commissioners of Forfeiture for southern New York. The 57,000 acres were born again as 287 separate properties auctioned by the revolutionary government, at a total benefit to New York State of over 220,000 pounds, more than twice the sum gained from the sale of the next largest confiscated estate.

Of the lots sold, all but eighteen went to farmers whose families had paid rent to their manor lords and ladies for generations. Historians of colonial land disputes hold that the new American government made a conscious decision to carve up the Philipse acreage (the average parcel weighed in at 170 acres) rather than locate a buyer for the entire package, which would have been easy. Orchestrating a public transfer of the grand property from the well-manicured hands of the aristocrats to the soil-tilling masses made a perfect symbol of democracy in action. Yet hundreds of the Philipses' cronies kept their land—dynasties such as the Livingstons, relations such as the Van Cortlandts—protected by their wartime allegiance to the Patriots. In fact, the choicest morsels from the Philipsburg spread were snatched up by two of the family's silver-spoon peers, the Upper Mills to Gerard Beekman and the Manor Hall to Cornelius Low.

The memory of the Philipse dynasty in the Hudson Valley gradually lost the taint acquired during the venomous Revolutionary era and acquired instead a time-mellowed patina with appropriate luster restored to men and women of so noble a station (no matter that their power was relatively recently invested and based in politics rather than ancient pedigree). An air of sadness clung to the Manor Hall. A British visitor passing through in 1787 recorded his impressions of what used to be "one of the finest Estates in the Country . . . The house of the Phillips family is a good old building with the best Gardens hot houses and Green houses in the country but the Person who now inhabits it being by no means in such affluent circumstances as Mr. Phillips it is not kept in so good a state as could be wished."

The Upper Mills fared better, as longtime tenants, now landowners, flourished. Today, a small part of the Upper Mills exists as a historical restoration, re-created as a farming-and-milling complex circa 1750, with guided tours, heirloom oxen, and a mill that grinds flour for purchase by visitors. Catherine's Dutch Reformed church stands across Broadway, none the worse for the wear after more than three hundred years. Though congregants have mainly gone to other Reformed churches nearby, the bronze bell Margaret imported from Holland still rings out for services every Sunday morning from June to September.

*F*or decades after the Revolution, Philipse Manor Hall passed from hand to hand until in 1813 a wealthy eccentric named Lemuel Wells, known for his limp and his gold-headed cane, bought the property at auction at the Tontine Coffee House on Wall Street in Manhattan. Wells and his wife, Rebecca, resided at the Manor Hall for twenty years. As the Hudson was the main trade artery in New York, and New York the wealthiest state in the nation, Rebecca and Lemuel became one of the more prosperous couples in an increasingly prosperous town.

Rebecca stayed on in the house following her husband's death in 1842, but because Lemuel left no will she could only watch as his estate was divided among sixteen heirs. Around 1850, the Manor Hall became a genteel boardinghouse, sheltering the families of some of the engineers who were just arriving in Yonkers to build the new railroad that ran straight along the Hudson's eastern edge. Yonkers grew glitzier. One measure of the boom was the success of a hotel called the Mansion House, which opened to great fanfare on Broadway, the rough trail whose paving Frederick II had overseen a century before. Nearby, St. John's Church, which had been paid for by Frederick II and erected by Frederick III, grew grander, with renovations that quadrupled its size.

The Manor Hall, in the mid-nineteenth century the center of a burgeoning, bustling town, nonetheless lay cushioned by a sort of pastoral feather bed. Boxwood hedges still screened the formal gardens, and the historic horse chestnuts still offered their cooling shade alongside the lines of locust trees. The mansion was lit by whale oil and candles and heated with isinglass-fronted Franklin stoves. One woman who lived there as a girl recalled that in front of the Manor Hall the Nepperhan was "surrounded with fields of wild violets that filled the air with perfume." Soon the Manor Hall was not the only mansion in the area, as magnates pitched their ornate villas on the bluffs above the river. A steamboat pier had been built at the shoreline where Margaret's trade sloops used to anchor, and many of the estimated forty thousand tourists who visited the picturesque Hudson every year found the old Philipse place a pretty and poignant draw as well.

The house was still the pivot around which Yonkers turned, even as the city grew urbanized with elevator factories, pencil mills, carpet mills, and other industries that enriched Gilded Age tycoons. The railroad separated the house from the shore and at the same time linked the countryside of former Philipse lands to the growing metropolis of New York City. Philipse Manor Hall passed from private hands in 1868 when Yonkers acquired the building for its first village hall. Without ado, eager politicos carved up the interior to make offices and ripped out the second-story walls to create an ornate Gothic courtroom. The city police department shared the site, and horse stables now stood west of the house's formal entry, where the sleighs of Joanna's party guests once pulled up to disgorge their passengers.

Today the building stands perfectly empty, inhabited only by light and ghosts, maintained by the New York State Parks Department, which does its best to keep the site intact given the usual budgetary limitations. From the house's west-facing windows you can still see the Palisades across the river, untouched by development, their towering crags and cool shadows identical to when Margaret first laid eyes upon them. But the rushing falls that set the Philipse millstones in motion have disappeared. Nineteenth-century civil engineers and health officials grew concerned about the factory waste deposited into the Nepperhan's millponds, and city planners who followed thought it wise to bury the polluted river under pavement. Since the 1920s, the tributary has coursed invisibly in an underground flume before gushing out from under a parking lot to join its waters with the Hudson. The orchards, deer park, and parterres of rare roses doted upon by generations of lords and ladies have shrunk to a postage-stamp lawn around the Manor Hall and a scrap of asphalt for visitor parking.

The Philipse heirlooms went to collectors at auction or disappeared once the family relocated to England—the now-priceless silver and rugs, the imported china, the generous *kasten,* the portraits that never did anyone justice. (Though a few of these treasures, like Copley's canvas of Mary, have since shown up on museum walls, to be gazed at by people who draw a blank at the surname Philipse.) And the ordinary things—the kid gloves, the birdcages, the chamber pots, the brass andirons chosen by the woman of the house to match the fireback in the west parlor—these long ago wended their way to the junk shops and swap meets where nameless old-time curiosities congregate in America.

Just a few blocks from the doorframe where Margaret leaned and smoked her long-stemmed pipe, the city of Yonkers has built a riverfront sculpture park to help revitalize the neighborhood around the old mansion. Here, alongside the shoreline where laden sloops once anchored, a local artist has installed a life-size likeness of a woman carved in limestone, clothed in seventeenth-century attire, and enthroned at the head of an imposing limestone table. Even were her name not chiseled in the rock frame surrounding her portrait, we would recognize the she-merchant by her implacable gaze, fixed upstream, sighting north along the shapeshifting surface of the Hudson as if she cannot wait for the season of trade to commence.

Acknowledgments

Without the keen mind and expert judgment of my wonderful editor Andrea Schulz, *The Women of the House* would not be the same book. She posed all the right questions at the right times. Thanks also to Sara Branch, who so carefully coaxed the book into its final form, to Lydia D'moch, who skillfully created its shape, and to Kelly Eismann, who provided such a beautiful skin.

To tell this story I drew on the words of people long in the past, and I am indebted to the staff at various research institutions for helping me access them. Most important were all those who provided support at the public library in Hastings-on-Hudson, especially Sue Feir, Janet Murphy, Kofi Addo-Nkum, Regina Kelly Houghteling and Dorothy Bingham.

The folks at Philipse Manor Hall State Historic Site in Yonkers opened their research files and bookshelves and made it possible for me to linger in the mansion to get a deeper sense of its private spaces. I thank Joanna Pessa, in particular, for her patience in touring the house with me, and for sharing her literate ideas about the Philipse family and their environment. The Manor Hall's Heather Iannucci and Lucille Sciacca also provided valuable support.

Cultural historian Alix Schnee, formerly with the Manor Hall, helped me clarify crucial issues in the family's past, in particular the mighty Margaret's course of action as the English influence began to cramp her style.

At Historic Hudson Valley, the central repository for all Philipse family papers, librarian Catalina Hannan provided guidance among the collection's treasures and allowed me access to vital maps, letters and paintings. She and Margaret Vetare welcomed me to set up a veritable home away from home in HHV's cozy yet incredibly extensive library.

The staff at various Philipse-related historic sites, including Philipsburg Manor and Sleepy Hollow Church in Sleepy Hollow, Van Cortlandt Manor in Croton, and the Van Cortlandt House Museum in the Bronx were unvaryingly informative and helpful.

I bear a special debt to all the people who had confidence in *The Women of the House* from the first. Josefa Mulaire read early pages, and continually reassured me that it is the imagining that makes history real. Thank you also to Sandra Robishaw, Lisa Senauke, Debbie Levitt, Christine Gilmore, Eric Saks, Gary Jacobsen, Bethany Pray, Barbara Feinberg, Suzanne Levine, Andy Zimmerman, and Peter Zimmerman. Betty and Steve Zimmerman contributed much more than required, as always, offering their interest, their constructive criticism, and their enthusiasm, and in general beefing up my confidence. Paige Wheeler gave the manuscript an early read and performed some welcome administrative tasks. Bart Plantenga made an invaluable long-distance connection.

Betsy Lerner gave me a gift that only an agent who shares a deep friendship with her author could offer: unbridled exhilaration over this project from the time it was only a creative germ. I don't know how she knew how much I would love to write this book, but she did. She made *The Women of the House* happen.

Without the unflagging interest of Gil Reavill—a man equally willing to go adventuring among the ghosts of Kingsbridge or to slog through comma choices—my immersion in this book would not have been nearly as enjoyable. He trusted I'd do right by my story, and had the generosity to take my work seriously when he was simultaneously consumed with his own. As for Maud Reavill, she showed me on a daily basis that she believed in this project, in me, and, sometimes, even in the fun of history.

Notes

Part One
One: Her New World

Among my most useful general sources regarding the settlement of New Netherland were Stokes, *Iconography of Manhattan Island, 1498–1909;* Van der Zee and Segal, *Sweet and Alien Land;* Van Rensselaer, *History of the City of New York;* Burrows and Wallace, *Gotham.* Three others were also informative: Cantwell and Wall, *Unearthing Gotham;* Barlow, *Forests and Wetlands of New York City;* and Augustyn and Cohen, *Manhattan in Maps 1527–1995.*

3 **At the bow** A number of works by Margaret's contemporaries describe entering New York Harbor, including: Benson, trans. and ed., *Peter Kalm's Travels in North America;* Dankers and Sluyter, *Journal of a Voyage to New York;* the selections included in Jameson, general ed., *Narratives of New Netherland* (especially that of explorer Peter de Vries); Van der Donck, *Description of the New Netherlands;* and Denton, *Brief Description of New York.* A good discussion of the watercraft of New Netherland can be found in Wilcoxen, "Ships and Work Boats of New Netherland, 1609–1674," *A Beautiful and Fruitful Place: Selected Rensselaerswijk Seminar Papers,* Nancy Zeller, ed.

3 **still a musket shot away** Peter Kalm in *Travels* (Benson, 330) describes the Hudson as a musket shot board. That would make a musket shot about a mile.

4 **Winebridge pier** Work Projects Administration, *Maritime History of New York,* 33; Stokes, *Iconography of Manhattan Island,* 2:214. Stokes writes that the pier "was begun by April 18, 1659, and was finished by July of the same year."

5 **Whether colonists** My description of the early settlement's commerce draws upon Rink, *Holland on the Hudson;* Condon, *New York Beginnings;* Earle, *Colonial Days;* Innes, *New Amsterdam and Its People;* Janvier, *In Old New York;* O'Callaghan, *History of New Netherland;* and Singleton, *Dutch New York.*

5 **"attends to it"** Nicolaes Van Wassenaer, "Historisch Verhael," in *Narratives of New Netherland,* ed. Jameson, 79.

5 **"starving times"** Linebaugh and Rediker, *Many-Headed Hydra,* 12.

6 **Eighteen languages** Isaac Jogues, "Novum Belgium," in *Narratives of New Netherland,* Jameson, 259.

7 **real Dutch community** "Letter from the Burgomasters of Amsterdam to Stuyvesant; Boys and Girls from the Almhouses Sent to New Netherland," Fernow, *Documents,* 325–26. For some, relocation was not a choice: In 1655 company officials plundered the almshouses of Amsterdam for farm labor, importing seventeen adolescent orphans.

7 **between 1630 and 1644** Rink, 147.

7 **consider a sampling of rosters** RootsWeb, Ship Passenger Lists 1623–1664, "Immigrants to New Netherland," http://www.rootsweb.com/~nycoloni/nnshdex.html.

8 **Her cousin, a well-to-do merchant** No evidence shows Margaret in America before the fall of 1659, which narrows the pool of possible ships to those that sailed in winter 1658–59 or spring 1659. It would be reasonable to expect Margaret to book space on the one that made the less arduous springtime crossing—the *Beaver,* say, which departed Amsterdam on April 25, 1659, and arrived at Manhattan two to three months later, in summer.

9 **an integral part** Maika, "Credit System of the Manhattan Merchants in the Seventeenth Century, Part I," *De Halve Maen,* 2.

10 **"making and receiving all Payments"** Mary Astell, "An Essay in Defence of the Female Sex," quoted in Schneir, "'She Merchants,'" 3.

10 **Census figures show** Biemer, *Women and Property,* 7.

Two: A Map of Manhattan

I have drawn mainly on these sources to suggest the social, work, and legal climate for the women of New Netherland and early New York: Biemer, *Women and Property* (especially the chapter "Margaret Hardenbroeck: Merchant, Shipowner, Supercargo"); Van Rensselaer, *The Goede Vrouw of Mana-ha-ta at Home and in Society 1609–1760;* Dexter, *Colonial Women of Affairs;* Agnew, "Silent Partners"; Narrett, "From Mutual Will to Male Prerogative: The Dutch Family and Anglicization in Colonial New York," *De Halve Maen* (Spring 1992); Shaw, "New Light from Old Sources: Finding Women in New Netherland's Courtrooms," *De Halve Maen* (Spring 2001); Van Zwieten, "'[O]n her woman's troth.'" For specific details on Margaret I relied primarily upon Biemer, *Women and Property;* Schneir, "'She Merchants'"; Randolph, "The Hardenbrook Family"; and Elliot, "Story of Philipsburg."

Notes

14 **"in the hands of Margriet Hardenbroeck"** February 28, 1660, statement, notarial archives no. 2735, September 19, 1659, Municipal Record Office, Amsterdam. Notary: Jan Hendricks. Leuven.

15 **"free agent"** Notarial Archives no. 2291 4:20–21, January 9, 1664, Municipal Record Office, Amsterdam. Notary: Jacob de Winter.

16 **quarter-mile-square Shangri-la** Cantwell and Wall, *Touring Gotham's Archaeological Past,* 6. Ruins dug up on Governors Island include hand-wrought nails and oak splinters from the base of a wind-powered sawmill dating to 1625.

17 **In black ink on vellum** Cortelyou was a brilliant Huguenot from Utrecht whose scholarly bent did not keep him from actively shaping the new colony (a few years before he drafted the 1660 map he founded New Utrecht on Long Island, building himself a home with a soaring view over the Narrows to Staten Island but a lousy commute to his Manhattan office).

18 **From Amsterdam, the directors** Letter from the Directors, December 24, 1660. Cited by Stokes.

18 **Margaret's arrival** Augustyn and Cohen (40) estimate the number of taverns as twenty-one; Rink (164) writes that two thousand would be a conservative guess for the total population in 1655, with thirty-five hundred the high end.

20 **sixteen million subjects** Cameron, ed. *Early Modern Europe,* 139. My principal sources for Holland's political and economic history were Boxer, *Dutch Seaborne Empire 1600–1800;* and Hooker, *History of Holland.* A rich understanding of the period comes from Haak, *Golden Age;* Van Deursen, *Plain Lives in a Golden Age;* and Schama, *Embarrassment of Riches.*

20 **seventeenth-century achievers** These include Hugo de Groot (1583–1645), the "Father of International Law"; philosopher Baruch Spinoza (1632–1677); painters Rembrandt van Rijn (1606–1669) and Jan Vermeer (1632–1675).

24 **The instigator found herself banished** Stokes, 2:261.

25 **less flashy property** Ibid., 270; Murphy, *Anthology of New Netherland,* 28.

25 **Oloff van Cortlandt** Stokes, 2:251.

26 **hogs rooted** Duffy, *History of Public Health,* 18–27. Over one hundred years later, in 1820, thirty thousand hogs roamed the streets of New York.

26 **Fort Amsterdam** One discarded brainchild of its designer was an eight-foot-deep moat encircling the base of the fortress.

27 **a painted ship's figurehead** According to Washington Irving in *Knickerbocker's History of New York* (68), presumably taking some romantic license.

27 **"in all quietness . . ."** Stokes, *Iconography of Manhattan Island,* 1:60.

27 **One early artist** "The City of New Amsterdam Located on the Island of Manhattan in New Netherland," 1650, watercolor, collection of Austria National Library. One exception occurred in 1641, when a slave named Manuel Gerrit (called the Giant) was sentenced to hang for killing another slave but received a pardon after the rope snapped under his weight. Gerrit would be freed by the Company three years later, one of the first slaves emancipated on Manhattan.

28 **pillory, iron, and post**   Records cited in Dunshee, *As You Pass By,* show that
two men received fifteen lashes apiece for the offense of "stealing fiddle strings."
Bastwick quoted in Earle, *Curious Punishments of Bygone Days.*

30 **"like a mole hill"**   Van der Donck et al., "Remonstrance of New Netherland,"
in *History of New Netherland,* ed., O'Callaghan, 1:303.

32 **Only slightly disemboweled**   Singleton, *Dutch New York,* 48.

33 **the gamut of features**   Frijhoff, "New Views on the Dutch Period of New
York," *De Halve Maen* 2 (Summer 1998): 22–34.

Three: Wild Diamonds

36 **a surfeit of vegetation**   Sources on the natural world the settlers encountered
include De Vries in Jameson, ed., *Narratives of New Netherland 1609–1664;* Van der
Donck; Wolley; *Two Years Journal in New York* (1678–1680); Hamilton in
Bridenbaugh, ed., *Gentleman's Progress;* Wood, *New England's Prospect;* and Miller,
"A Description of the Province and City of New York; With Plans of the City and
Several Forts as they Existed in the Year 1695" in Jaray, ed., *Historic Chronicles of
New Amsterdam.* Van der Donck noted a celebrated event of March 1647, when a
whale was spotted riding the salt tides of the Hudson many miles upriver, where it
washed ashore near "the great Chahoos falls" and was promptly set upon by
colonists for its oil. Embarked on a sloop to Jamestown, Virginia, De Vries was
shocked to see a sturgeon eight feet long spring spontaneously out of the Delaware
River onto the deck of his sloop.

39 **Above the palisade**   Sources for life on northern Manhattan include Dunshee and
Stokes.

40 **a thriving, complex society**   When European mariners first stumbled upon
the future New York Harbor in the 1500s, as many as fifty thousand people
crowded Lenapehoking, an estimate supported by the eighty habitation sites and
two dozen planting fields archaeologists have found within the five boroughs of
New York alone. Burrows and Wallace, 5. Some scholars put the figure lower, at
fifteen to thirty-two thousand. Otto, *New Netherland Frontier;* Brawer, gen. ed.,
Many Trails; Kraft, ed., *Archaeology and Ethnohistory.* Material on trails and
settlements drawn from Bolton, *Indian Paths;* Skinner, *Indians of Manhattan;* and
Grumet, "'We Are Not So Great Fools.'"

40 **"island of hills"**   Homberger and Hudson, 17. Linguists have alternatively traced
the name as "the place we got drunk."

41 **"fascinating"**   Van der Donck, 73.

41 **seven epidemics**   Kraft, "Indians of the Lower Hudson Valley at the time of
European Contact," *Archaeology and Ethnohistory,* 212. Dutch pioneeers could not
help but see tragedy as windfall. Jogues praised the "lands fit for use, deserted by
the savages, who formerly had fields here." Jameson, ed., *Narratives of New
Netherland,* 261.

42 **Some inland Lenapes traveled** Useful in this discussion of early Dutch-Indian trade: Bachman, *Peltries or Plantations;* Brawer, *Many Trails;* and Brasser, "Early Indian-European Contacts," in Bruce Trigger, ed., *Handbook of North American Indians:* vol. 15, *Northeast* (Washington: Smithsonian Institution, 1978). Trade values estimated in Van Rensselaer, *History of the City of New York.*

44 **Soon enough, the Dutch Reformed Church** Fisher, *Men, Women & Manners.*

45 **"should not be driven away by force"** F. C. Wieder, *De Stichting van New York,* quoted in Van der Zee, 2.

45 **Good ships were worth a fortune** William David Barry and Frances W. Peabody, *Tate House: Crown of the Maine Mast Trade* (Portland, ME: National Society of Colonial Dames of America in the State of Maine, 1982); John T. Faris, *When America Was Young* (New York: Harper & Brothers, 1925), 305.

47 **annual exports of hides** Louis Jordan, "Money Substitutes in New Netherland and Early New York: The Beaver Pelt," Department of Special Collections, University of Notre Dame Libraries, http://www.coins.nd.edu/ColCoin/ColCoinIntros/NNBeaver.html.

47 **"proper for muffs"** Van der Donck, 43.

48 **England's charismatic style-setter** In his diary, Pepys describes Charles inventing the waistcoat in 1666, pairing it with knee breeches and a long jacket—the original three-piece suit.

Four: A Wedding, a Child, and a Funeral on the Ditch

51 **Diversions filled the entire town** Singleton, *Dutch New York;* Earle, *Colonial Days;* Cor Snabel, "Life in Amsterdam," The Olive Tree Genealogy, http://www.olivetreegenealogy.com/nn/. According to Judd in "Thanksgiving in New York," a new holiday followed the settlers to America—a day of thanksgiving that celebrated the deliverance of the Dutch from Spanish clutches on October 3.

53 **Pieter made a trip** Biemer, *Women and Property,* 34. Information on marriage and other social customs furnished in Van Zwieten, "'[O]n her woman's troth.'"

56 **constructed of clapboards** Meeske, *Hudson Valley Dutch,* 200–1.

57 **Pregnant Dutch women** For Dutch birthing practices and rituals I relied primarily upon Van Zwieten; Paul Zumthor, *Daily Life in Rembrandt's Holland* (New York: Macmillan, 1963); and Earle, *Colonial Days.* For details in America, the most useful were Wertz and Wertz, *Lying-In;* Scholten, *Childbearing in American Society;* and Phyllis Putman, "Child-Bed Linen," Colonial Williamsburg, http://www.history.org/history/clothing/milliner/childbed.html.

58 **Next to the director-general** Van Zwieten in "'[O]n her woman's troth'" (11) cites two other midwives between the 1630s and 1660s in New Netherland: Lysbet Dircks and Hillegond Joris.

60 **"depart alone to a secluded place"** Van der Donck, 84.

61 **50 percent** Van Zwieten, 12.

61 **Tappan, New Jersey** Ibid.

61 **"joyful mothers"** Scholten, 9, with details about clothing on 15–16; maternity garb also discussed in Singleton. Other costume details from Tom Tierney, *Colonial and Early American Fashions* (Mineola, NY: Dover, 1999); Douglas Gorsline, *What People Wore: 1,800 Illustrations from Ancient Times to the Early Twentieth Century* (Mineola, NY: Dover, 1980); and Warwick, Pitz, and Wyckoff, *Early American Dress*.

62 **passage from Amsterdam** Indenture of Abel Hardenbroeck, January 6, 1659, notarial archives 1359 fol. 6., Municipal Record Office, Amsterdam. Notary: Hendrick Schaeff; Randolph, "The Hardenbrook Family," 130.

67 **"Hot sickness"** Stokes, vol. 2; Duffy, *History of Public Health,* 19.

67 **Claus van Elslandt** Details on funeral customs and dress drawn from Stokes, *Iconography of Manhattan Island,* 2:222; Earle, *Colonial Days,* 293–312; Wilcoxen, *Seventeenth Century Albany,* 122–23; Warwick, Pitz, and Wyckoff, 139–141; and Blackburn and Piwonka, *Remembrance of Patria,* 205.

68 **much distraction** Stokes, 2:343.

70 **four substantial Manhattan properties** Biemer, *Women and Property,* 118. Pieter had owned lots 10, 19, and 34 in Block C, and lot number 2, Block F, in Manhattan.

71 **She had branched out** Record of Margaret's agreement with Daniel des Messieres to pay the proceeds of a cask of oil and some pins. March 2, 1660, notarial archives 2735, Municipal Record Office, Amsterdam. Notary: Jan Hendricksz. Leuven.

71 **Of all the shops along the Strand** Information on the coopering trade from Sloane, *American Barns and Covered Bridges* (Mineola, NY: Dover, 1954), 22–23; "A Squandered Inheritance" in John T. Faris, *When America Was Young* (New York: Harper & Brothers, 1925); Maika, "Commerce and Community," 470–72.

Five: Education of a She-Merchant

73 **As a child in Amsterdam** Details on education in Holland and New York primarily drawn from Kilpatrick, *Dutch Schools of New Netherland;* Singleton, *Dutch New York;* and James R. Tanis, "Growing Knowledge in New Netherland," *De Halve Maen* 63, no. 3 (September 1989), 1.

74 **"in all sorts of languages"** Van Zwieten, 7.

76 **"not a cobbler in these parts"** Boxer, 155.

76 **Amsterdam's bridegrooms** Willem Rabbelier and Cor Snabel, "The Crossing: Routes and Duration," The Olive Tree Genealogy Pages, http://www.olivetreegenealogy.com/nn/mm_4.shtml.

76 **the percentage for Dutch women** Boxer, 155; Van Zwieten, 9.

77 **"about 200 scholars"** Hamilton in Bridenbaugh, ed., *Gentleman's Progress,* 78.

79 **"explosively mercantile"** Rink, 86.

79 **"more fond of and delighted with lasciviousness"** John Ray, quoted in Schama, 402.

Notes

80   **"women are fit for the household"**   Quoted in Schama, 419.

81   *manus*   Dutch marital provisions are elucidated in Biemer, *Women and Property*, 1–2.

83   **"a weak creature"**   1562 Homily on Marriage quoted in Voorhees, "'how ther poor wives do.'"

Six: A Marriage of Love and Trade

85   **he had outfitted a good-size sloop**   "Resolution to charter to Frederick Philipse, late the director's carpenter . . ." Sept. 20, 1660. E. B. O'Callaghan, ed., *Calendar of Historical Manuscripts in the Office of the Secretary of States* (Albany: Weed, Parsons, 1865–66), 218.

86   **"Fredrick Flipsen, Carpenter"**   Biemer, *Women and Property*, 36.

88   **The Orphanmasters was a concept**   A selection of these canvases can be seen in Haak.

89   **"keep their eyes open"**   Berthold Fernow, *The Minutes of the Orphanmasters of New Amsterdam 1655–1663* (New York: Francis P. Harper, 1902), 222. The saga of Eva's adoption is related on 222–23, 225–26, 226–28; in Elliott, 38; and in Biemer, *Women and Property*, 36.

91   **In 1662 Johannes de Witt alleged**   "Johannes de Witt, pltf.v/s Margriet Hardenbroeck, deft., January 17, 1662," Fernow, ed. *Records of New Amsterdam*, 4:13–14.

91   **In January 1664**   January 9, 1664, notarial archives no. 2291, 4:20–21, Municipal Record Office, Amsterdam. Notary: Jacob de Winter.

92   **Cornelia Lubbetse**   Dexter, 103–4.

92   **Frederick was dogged**   "Complaint against Frederick Philipse, Nicholas Myer, and others, for driving wagons loaded with grain from Wiltwyck to the Rondout, without a convoy," December 8, 1663, Fernow, *Documents*, 312–13.

93   **"dead bodies of men"**   Van der Zee, 404–5.

95   **Adolph received an appointment**   Van Rensselaer, *History of the City of New York*, 420; Randolph, 128.

96   **a property Frederick owned**   Stokes, 2:249.

97   **the all-important *kas***   Denton, "American Kas," *De Halve Maen* (13); Kenney, Safford, and Vincent, *American Kasten*.

98   **Two other staple chores, baking and brewing**   Schneir, 15.

99   **One colonist's far-from-atypical inventory**   Fisher, *Men, Women & Manners*.

Seven: The Superior Authority Over Both Ship and Cargo

101   **They were militarily outnumbered**   O'Callaghan, *History of New Netherland*, 2:529.

103   **It seemed Stuyvesant**   Later, after having been cashiered by the West India

Company, he delivered a bitter explication of the saga to the States-General, the governing body of the United Provinces. In his account, he decried the "disrespectful [. . .] treatment" he had received at the hands of the citizenry, and the "general discontent and unwillingness to assist in defending the place" that he had been burdened with.

104 **The slave ship** *Gideon* The cargo represented about 3 percent of the colony's population, almost half the number of Africans already enslaved there.

106 **With one exception** Still, though, when that conflict ended, the States-General deemed the fur-rich wilderness so negligible in value that they gave it up a second time.

106 **a most attractive profit-generating venue** Doris C. Quinn, "Theft of the Manhattans," *De Halve Maen* (Summer 1993).

108 **experienced traders such as Margaret Hardenbroeck** A useful discussion of these circumstances in Biemer, *Women and Property;* and Schneir.

108 **Jointly with her peers** "Petition of Oliver Stuyvesant Van Cortlandt, and others," Brodhead, *Documents,* 3:178–79.

109 **Margaret and Frederick took advantage** "Trade pass for Frederick Phillips of New York or his wife Margarett, to Albany to traffic with the Indians or others," March 21, 1664/5. Colonial Records (Albany, 1899), General Entries, vol. 1, 1664–65.

109 **Changes in currency valuation** Wolley, 58. Analysis of sewan inflation by D. T. Valentine, *Manual of the Corporation of the City of New York for 1855* (New York: Common Council, 1841–1870), 660.

111 **Governor Andros, appointed in 1674** He received his appointment just after the Dutch waged a fifteen-month war to win back what they briefly renamed "New Orange" (and dropped it yet again in favor of Surinam).

111 **The choice was politic** Bonomi, *A Factious People,* 60.

112 **gradual elimination of Manhattan's independent she-merchants** In "Criminal Law" (76), Biemer puts the number of women operating as traders in New Amsterdam from 1653 to 1663 at 134, while after the British takeover the city of New York could claim just 43 (the identification of commercial women was by no means absolute, however, as some now operated under the mantle of their husbands). The number of female proprietors of businesses (tapsters, bakers, etc.) also declined, from fifty in 1653–63 to seventeen in 1664–74. In Albany the slide was even more precipitous, from forty-six traders in 1654 to zero in 1695–1700. And by 1695–1700 only three women had shops open in Albany, down by ten from the decade before the English came.

112 **power of attorney** In 1670 Jacques Cousseau, who declared himself "to be well & truly indebted unto Mdme Margriet Hardenbrook wife and attorney acting in absence of her husband Mr. Frederick Philipsen," swore in court that the large sum of guilders (eighteen hundred) he had owed Philipse since 1668 would be paid just as soon as the *Fort Albany* arrived from Amsterdam with that same value in

"ossenburgh." Cousseau offered as collateral three houses he owned—one on upscale Pearl Street and another on the Heerengracht, along with a millhouse and horse mill, and a promissory note for 3,520 pounds of Virginia tobacco one Pieter Aldricx owed Cousseau. August 29, 1670, notarial archives no. 2291, Municipal Record Office, Amsterdam. This was a big-ticket commercial transaction, one in which a husband, Philipse, gave his wife his power of attorney, a right that was still allowed these six years after the coverture-crazed British took control of Manhattan and its female occupants.

113 **"in hopes to make her free"**   Sir John Werden to Governor Andros, St. James. May 7, 1677. Brodhead, *Documents,* 3:246–47.

113 **the premier fur market**   Bachman, *Peltries or Plantations,* 23.

113 **Each spouse had a namesake ship**   Biemer in *Women and Property* (120) notes fifteen, perhaps omitting some co-owned vessels. Historic Hudson Valley researchers have estimated twenty. "Ships Belonging to Frederick Philipse," unpublished research report, Historic Hudson Valley.

113 **A typical bill of lading**   Bonomi, 61.

115 **"very prophane and godless language"**   "Ordinary session held in Albany on Tuesday, the 24th of August Anno Domini 1675." A. J. F. Van Laer, ed., *Minutes of the Court of Albany, Rensselaerswyck and Schenectady, 1675–1680,* vol. 2 (Albany, 1928), 9–12.

118 **Passenger Jasper Danckaerts**   Dankers and Sluyter, *Journal of a Voyage to New York.* The work was discovered nearly two centuries after it was written by an antiquarian bookseller in Amsterdam.

119 **eighty-eight different types of goods**   Biemer, *Women and Property,* 95.

Eight: The House Margaret Built

127 **Margaret launched an epic buying spree**   Biemer, *Women and Property,* 42. Shonnard and Spooner, Westchester County, "The Royal Charter," C. W. Bolton, *History of the County of Westchester,* ed. Robert Bolton (New York: A. S. Gould, 1848); Allison, *History of Yonkers,* 55; Scharf, *History of Westchester County,* 1:72, 96.

127 **Colen Donck**   The concept of patroonship was a feudal throwback that the West India Company had instituted early on in the life of the colony to encourage migration, stipulating that any Company member who could bring over fifty people age fifteen and older as settlers would receive sixteen miles of prime riverfront real estate, extending "as far inland as the situation would permit." With the patroonships, the Directors thought to entice citizens to invest in huge parcels of land that they would settle with colonists at their own expense, thus achieving both the Company's objective of populating New Netherland and the patroon's of getting ever richer. Van der Donck had himself been land hungry, in that respect differing from so many of the settlers in New Netherland. Even most of the

patroonships lay idle, since most colonists lived to trade and could make a bigger killing buying and selling in New York than managing farmland.

128 **Frederick Philipse signed a deed** Haley, "Philipsburg Manor Tricentennial, 1693–1993," 11, quoting from Robert Bolton, *The History of the Several Towns, Manors, and Patents of the County of Westchester. . . . from Its First Settlement to the Present Time,* 2 vols. (New York: Chas. F. Roper, 1881), 1:507.

129 **whose traditional capital** The recent history of these lands must have thrown a shadow over the new owners' businesslike deed signings. As early as 1626, Company soldiers ambushed a Weckquaesgeck uncle and nephew as they trekked into New Amsterdam with a load of furs. In 1642, an expedition of eighty soldiers fought the Lenapes of the Bronx and Westchester on Kieft's orders. The next year the Weckquaesgecks seized two Manhattan-bound boats and killed the family of Anne Hutchinson, actions that English soldier John Underhill avenged with a midnight massacre of five hundred native men, women, and children. A pattern of mistrust had long since been ingrained.

129 **letter from one Frenchman** Siere de Saurel to Intendant Duchesnau of New France, received November 14, 1679. Brodhead, *Documents,* 9:138–39.

130 **"They desire as before from Mr. Philips"** Council minutes, "Examination of Westchester Indians as to their Intention to Join King Philip." March 29, 1676. Fernow, *Documents Relating to the History,* 494–95.

131 **Probably the most reliable computation** Vetare, *Philipsburg Manor Upper Mills,* 60.

132 **And so in 1682** Maika, "Philipse Family Commerce." A later voyage, of the *Beaver* in 1686, would haul an equal load of bricks along with six large cast iron firebacks for chimneys, another pair of millstones, and twelve dozen corn scythes.

137 **"English milk"** Henry, *Travels and Adventures.*

138 **a great comet** "The Commissaries of Albany to Capt Brockholes," January 1, 1681, *Documentary History of New-York,* 3:882. It was not only sky signs that held meaning for seventeenth-century minds—the most and least sophisticated alike. After explorers began to bring new biological specimens from Africa and the Far East back to Europe, Renaissance churches integrated the protective strength of these curiosities with the traditional Christian teachings. It was not uncommon for kneeling parishioners to take inspiration from a menagerie of marvels suspended by cords from the Gothic arches: ostrich eggs, elephant tusks, whale ribs, and toothy, desiccated crocodiles. Lorraine Daston and Katherine Park, *Wonders and the Order of Nature, 1150–1750* (New York: Zone Books, 1998), 52.

138 **about seven hundred year-round residents** Miller, "Description of the Province." For an understanding of colonial Albany, I have drawn on: Blackburn and Piwonka; Wilcoxen; Venema; Armour, *Merchants of Albany;* and Merwick, *Possessing Albany.*

139 **"In the early days"** Fisher, *Men, Women & Manners.*

140 **a fine house on Yonkheer Street** "Conveyance of a house and lot from Pieter

Meuse and Volkje Pieterse to Frederick Phillipse. . . ." *Fort Orange Records 1656–1678,* 189. Adolph and Maria relocated to Albany the year their daughter received her trading pass, which we know because frugal Frederick was quick to appeal the town's tax on the house occupied by Hardenbroeck, saying that Hardenbroeck served as "merely his agent," and since Frederick already paid taxes on his estate in New York he prays that the tax "might be taken off." "Petition. Frederick Philips, complaining. . . ." September 27, 1674 (Eng. MSS. 23:396).

141 **eight thousand gallons**   Norton, *Fur Trade,* 32.

144 **Adolph had already begun to make his way**   Elliott, 139.

144 **"free merchant"**   Notarial archives no. 2327, 267, Municipal Record Office, Amsterdam.

144 **enlarge the scope**   New York Colonial MSS, New York State Library, 38:71

144 **the St. Mary**   Ibid., 38:135.

144 *Land of Promise*   O'Callaghan, 158.

144 **"cruise along the coast"**   Commission from Jacob Leisler, July 23, 1690. Eng. MSS. 36:142.

145 **She even managed a brewery**   Peter R. Christoph, "Worthy, Virtuous Juffrouw Maria van Rensselaer," *De Halve Maen* 70:2 (Summer 1997), 25. Maria wanted, Jeremias wrote his mother, to re-create the world of her childhood: "I have taken up brewing, and this for the sake of my wife, as in her father's house she always had the management thereof, to wit, the disposal of the beer and helping to find customers for it."

Nine: A Surfeit of Sugar

To discuss colonial Barbados I drew most heavily upon Dunn, *Sugar and Slaves;* Beckles, *History of Barbados;* Beckles, *Natural Rebels.* Of the many histories of the slave trade, I found Thomas, *The Slave Trade,* the most useful.

150 **"This island"**   Henry Whistler, 1655, quoted in Dunn, 77.

151 **"a life built round cane waste"**   Thomas, 137.

152 **By 1660 sugar profits**   Beckles, *History of Barbados.*

155 **"parrot from the Spanish coast"**   *Valentine's Manual,* 1857, 512.

157 **On English slave ships**   Rawley, *Transatlantic Slave Trade,* 300.

157 **Even the female captives**   Beckles, *Natural Rebels,* 155.

158 **The West India Company, meanwhile**   From the 1630s to the 1650s, Holland was "unquestionably the dominant European slave trader in Africa." Herbert S. Klein, *The Atlantic Slave Trade* (Cambridge: Cambridge University Press, 1999), 76–77.

159 **the Charles, sailing from the West Indies**   The saga of the first slaves at Philipsburg is related in Haley, "Philipsburg Manor Tricentennial," 17; and Vetare, 21.

160 **Frederick chose English names** Frederick Philipse I, will dated October 26, 1700. *Abstracts of Wills on File in the Surrogate's Office, City of New York,* vol. 1, *1665–1707,* New York (County) Surrogate's Court, 369.

162 **In November 1690 she and Frederick** Indenture between Frederick Philipse and Jacobus van Cortlandt, November 29, 1690, ID 908, The Papers of John Jay, New York Historical Society.

163 **the plea of her ailing sister** Christoph, "Worthy, Virtuous Juffrouw," 27.

163 **Her inheritance upon her husband's death** *Abstracts of Wills,* vol. 2, *1708–1728,* 396. She was the sole beneficiary of the bountiful-sounding "all my lands and tenements in New York, [. . .] and all my goods and chattels," John Dervall cited in his will.

164 **Manhattan's designated** *aanspreckers* Earle, *Colonial Days,* 293.

Part Two
Ten: The Church of Catherine

In describing an earlier Madagascar I have drawn upon Frans Lanting, *Madagascar: A World Out of Time* (New York: Aperture, 1990); Peter Tyson, *The Eighth Continent: Life, Death, and Discovery in the Lost World of Madagascar* (New York: William Morrow, 2000).

171 **Frederick had even suffered prosecution** Maika, *Commerce and Community,* 474.

173 **Nuptial festivities in 1690s New York** Earle, *Colonial Days,* 56–69.

176 **"sun-expelling mask"** Earle, *Costume of Colonial Times.*

179 **pirate fever** Some of these details come from Carse, *Ports of Call.* Other sources on pirates: Rogozinski, *Honor Among Thieves;* Cordingly, *Under the Black Flag;* and Linebaugh and Rediker, *Many-Headed Hydra.*

181 **"burned in parts that for decency he will not refer to"** Cordingly, 131–32.

182 **Frederick sent the** *Charles* **on a round-trip voyage** Judd, "Frederick Philipse and the Madagascar Trade," 358.

183 **The devotion to rum was not limited** Thomas, 328.

183 **His vessels made their way out** They blended in with the rest of the burgeoning harbor traffic: Whereas in 1678 there were three ships, eight sloops, and seven boats using the port of Manhattan, sixteen years later there were sixty ships, sixty-three sloops, and forty boats. Carse, *Ports of Call,* 153.

184 **"a young negro wench and child of 9 months"** A. J. Williams-Myers, "Re-examining Slavery in New York," *New York Archives Magazine* (Winter 2002), http://www.nysarchivestrust.org/apt/archivesmag/archivesmag_past.shtml#Winter2002Feature.

184 **Frederick brokered one last deal** His ship, the *New York Merchant,* returned from Madagascar to New York in 1698 with its cargo. Under cover of darkness, his other ship, the *Frederick,* met the *New York Merchant* at sea, off the coast of

Delaware. There, all East India merchandise was transferred from the *Merchant* to the *Frederick*, and the *Frederick* continued on to Hamburg. With its slaves still aboard, the *Merchant* continued to New York Harbor. In June 1698 the *Frederick* was seized in Hamburg for violating the Acts of Trade and Navigation. Its goods were forfeited to the Crown. Soon after, Burgess stood accused of piracy.

185 **"With hearty greetings"**   Christoph, "Worthy, Virtuous Juffrouw," 27.

188 **The time was ripe**   The manor grant for Philipsburg included the right of "advowson," or patronage, for a local church and minister.

190 **"cold cement"**   *Building with Stone and Brick* (Buffalo, NY: Firefly Books, 1998).

194 **a single parish on the island**   Yellow fever "decimated the Barbados population in the 1690s," according to Dunn, 103.

194 **"Many who had begun"**   Ibid., 77.

195 **had been a far-off worry**   Bob Arnebeck, "A Short History of Yellow Fever in the U.S."; Duffy, *History of Public Health,* 35.

197 **Frederick bequeathed to his stepdaughter**   Frederick Philipse I, will.

Eleven: Not Doubting of Her Care

202 **In 1707 a baptismal record**   "New Amsterdam & New York Reformed Dutch Church Baptisms 1707," The Olive Tree Genealogy, http://www.olivetreegenealogy.com/nn/church/rdcbapt_1707.shtml.

202 **That was the last**   Her return announces itself in November 1716, when "Widdow Philips" appears on a handwritten "Account of Docking of Sundry Boats and Vesells Comeing into the Dock & Slips of the City of New York," a record that lists her as having paid the dockage for two boats. "An Account of Docking and Sundry Boats . . . ," New York Historical Society.

207 **In 1719, the year he turned twenty-one**   Minutes of the Common Council, quoted in Elliott, 190.

211 **"all sorts of House work; she can Brew, Bake, boyle soaft Soap"**   "An advertisement in the New York Weekly Journal Offering a Slave for Sale (1734)," *A Teacher's Guide to Understanding the Enslaved World in New York and Philipsburg* (Tarrytown, NY: Historic Hudson Valley).

212 **The population constantly expanded**   An estimated sixty-eight hundred Africans arrived as slaves in New York between 1700 and 1774. Williams-Myers, *Long Hammering,* 21.

212 **The races occupied extremely close quarters**   Apthecker, *American Negro Slave Revolts,* 172–73.

213 **The southern rice-indigo-tobacco-growing colonies**   Ibid., 164.

213 **The system of slavery**   African Burial Ground, http://www.africanburialground.com.

213 **"They had resolved to revenge themselves"**   Letter from Governor Robert Hunter, June 23, 1712, in Brodhead, *Documents,* 5:341–42.

213 **As of 1712, New York's slaves could not travel independently** Williams-Myers, 44.

214 **The ads demonstrate why affluent** Hodges and Brown, eds. *"Pretends to Be Free,"* 7.

215 **its share of runaways** Haley, "The Slaves of Philipsburg Manor," 6.

220 **"Molly and Sarah My Indian or Mustee Slaves"** Catherine Phillipse, will dated January 7, 1730. Copy at Historic Hudson Valley, Tarrytown, NY.

Twelve: Fashion Babies

To understand the fashions, cosmetics, and food and drink of the 1740s I drew upon Warwick, Pitz, and Wyckoff; Meeske; Crowley, *The Invention of Comfort;* Belden, *The Festive Tradition;* Baumgarten, *What Clothes Reveal;* Singleton, *Social New York Under the Georges 1714–1776;* Gunn, *The Artificial Face.*

227 **the median number of windows** Crowley, 103.

230 **"some are so furious"** Knight, *Journal of Madam Knight,* 31.

231 **"gynecandrical dancing"** Salinger, *Taverns and Drinking,* 32.

232 **"Publick and Common General Highway"** *Colonial Laws of New York from the Year 1664 to the Revolution* (Albany, NY: J. B. Lyon, state printer, 1896), 1:533.

233 **"old rule"** Belden, 91.

234 **Later would come yellow ocher** Meeske, 331; Frank S. Welsh, "The Early American Palette," in Moss, ed., *Paint in America,* 75.

238 **"A Sincere heart"** Willett and Cunnington, *History of Underclothes,* 72.

241 **Benjamin Franklin, for example** engraving of Benjamin Franklin by John Martin Will, after a drawing by Charles Nicolas Cochin, France, 1777.

242 **Gentlemen cast their own pewter buttons** Singleton, *Social New York,* 190.

242 **"We see they are Friends to Men"** Robert Campbell, *The London Tradesman* (1747; repr., New York: August M. Kelley, 1969).

242 **one popular rustic amusement** Willett and Cunnington, 70.

243 **"such sleek materials as silk and satin"** Warwick, Pitz, Wyckoff, 247.

Thirteen: A Hard Winter and Hell

245 **an elegant, teeth-clattering box** Road conditions in eighteenth-century America were not so far removed from those Cromwell availed himself of in seventeenth-century England when he captured eight hundred horses that were stuck in highway mud.

246 **liveried to the nines** Baumgarten, 128.

247 **one peppercorn** Common Council resolution, October 1, 1737, in Stokes, 4:537.

247 **A map drafted in 1735** Plan of the city of New York in the year 1735, the New York Public Library, I. N. Phelps Stokes Collection.

248 **"thirty or forty gentlemen and ladies meet"** Burnaby, *Travels Through the Middle Settlements.*

249 **already known as fireworks** In 1749 George Frederick Handel composed the piece "Musick for the Royal Fireworks" to accompany the display over the Thames to celebrate England's Treaty of Aix-la-Chapelle (the performance was cut short due to an outbreak of fire).

249 **Shakespearean tragedy would arrive in 1750** T. Allston Brown, "The First Play Acted in New York," *A History of the New York Stage: From the First Performance in 1732 to 1901* (New York: Dodd Mead, 1903); Arthur Hornblow, *A History of the Theatre in America*, vol. 1 (Philadelphia: J. B. Lippincott, 1919).

249 **Clever mechanical devices** Gottesman, *Arts and Crafts in New York*, 374–94; Singleton, *Social New York*, 320–21.

250 **An enthusiasm for horse racing** James Sullivan, ed., *The History of New York State 1523–1927*, vol. 2, chap. 3, pt. 6 (Chicago: Lewis Historical Publishing, 1927).

250 **"veal, beef stakes"** Bridenbaugh, 48.

254 **The independent-minded *Weekly Journal*** In early 1734, in elections for the city Common Council, the Government Party again found itself squarely defeated. The electorate showed an unprecedented distaste for the standard squad of merchant-princes, and installed a council whose ranks included three bakers, a bricklayer, a painter, and a bolter.

256 **the Hard Winter** Ludlum, *Early American Winters*, 49; detail about herds of deer from Benson, *Peter Kalm's Travels in North America.*

258 **"the Air that he gave us"** James MacSparran, "A sermon preached at Naraganset, R.I.," quoted in Ludlum, 49.

259 **slave uprisings in the other English colonies** Apthecker, *American Negro Slave Revolts.*

264 **"This crime is of so shocking a nature"** Quotes and details from Daniel Horsmanden, *Journal of the Proceedings in the Detection of the Conspiracy Formed by Some White People. . . .* (New York, 1744), as cited in Lepore, *New York Burning*, and T. J. Davis, *Rumor of Revolt.*

271 **"at the House of the late Adolph Philipse"** Advertisement, *New-York Weekly Post-Boy*, April 9, 1750.

273 **Frederick did inscribe a will** Frederick Philipse II, will, liber 18, 1, *Abstracts of Wills*, vol. 4, *1744–1753*, 355–59.

Fourteen: A Castle on the Heights

277 **"gay and numerous assembly"** Singleton, *Social New York*, 306–7.

279 **"a fine fat Ox"** Tieck, *Riverdale, Kingsbridge, Spuyten Duyvil.*

285 **"to prevent the killing and destroying of partridges"** Bielinski, *An American Loyalist*, 15.

286 **"well tempered, amiable man"**   Judd, "Frederick Philipse III of Westchester
County: A Reluctant Loyalist," in East and Judd, eds. *Loyalist Americans,* 30.

290 **"Timothy Scandal Adjutant"**   Timothy Scandal Adjutant, "A Return of the
State of Capt. Polly Phillips's Dependant Company, with the Kill'd Wounded,
Deserted, and Discharg'd &c, during the Campaigns 1756 & 1756 [*sic*]," December
25, 1756, LO 6475, Huntington Library, San Marino, California. Early relationship
of Morris and Mary drawn from Hayes, "Liberty Belles: Home Town Girls," *The
Westchester Historian* (Winter 1999) and Grieff, *Morris-Jumel Mansion.*

293 **The wedding party's members arrived at the Manor Hall by sleigh**
Philipse Manor Hall, 117.

294 **The ancient pedigree declared by such a crest**   Schriek, *Philipse Jewel.*

297 **"The only principle of Life"**   Singleton, *Social New York,* 314–15.

301 **"laid out a great deal in stone fence"**   Roger Morris, as quoted in Grieff, 100.

301 **"five hundred messuages"**   Indenture Quintipartite, Philip Philipse, Beverly
and Susannah Robinson, Mary Philipse, William Alexander, and Thomas Jones,
June 29, 1753, in William Pelletreau, *History of Putnam County, New York*
(Philadelphia: W. W. Preston, 1886), 45.

302 **the native population on both sides**   Pelletreau, *History of Putnam County.*

303 **the Philipses had seized**   Some tenants who had formerly held Philipse leases
now sided with the Wappingers because those terms were more favorable.

Fifteen: Fire in the Sky

309 **"when he supposed he was not noticed"**   Account of George Hewes, 1773,
in James Hawkes, *A Retrospect of the Boston Tea Party: With a Memoir of George R. T.
Hewes* (New York: S. S. Bliss, 1834).

310 **When a Massachusetts woman bore a son**   Berkin, *First Generations,* 177.

314 **"Copley's Performance"**   Roger Morris to Mary Philipse Morris, September
2, 1775, *Letters.*

316 **"The Steeple which was 140 Feet high"**   Shelton, *Jumel Mansion,* 46.

318 **"Would to God"**   Roger Morris to Mary Philipse Morris, October 1775,
Letters, 28–29.

319 **"come home to my lodgings"**   Ibid., March 4, 1777, 102.

319 **"scorbutic complaints"**   Ibid., August 1, 1776, 72.

319 **"too womanish"**   Ibid., July 17, 1775, 15.

319 **"is certainly my dear a fine girl"**   Ibid., December 23, 1775, 43.

320 **"the other so much inflamed as to make me very cautious"**   Bielinski, 22.

320 **"A few good Lemons would be verry acceptable"**   Ibid., 41.

321 **Battle of Brooklyn**   The British force included two men-of-war, two dozen
frigates, four hundred transports, thirty-two thousand troops, and thirteen thousand
seamen, in contrast with twenty-three thousand ill-equipped American soldiers
with inadequate equipment and training and not one ship.

323 **"treated all here without discrimination"** and **"Many helpless women . . ."** Crary, in *Loyalist Americans,* 17.

324 **"small and pretty feet"** *Letters of Hessian Officers* (New York: William L. Stone, 1891).

324 **"unmercifully pillaged"** Stephen Kemble, diary entry from November 2, 1776, quoted in Hall, *Philipse Manor Hall,* 153.

324 **"stripping the women"** Crary, *Price of Loyalty,* 177.

325 **"The Misfortunes of War"** and **"Neutrals in civil country"** East and Judd, *Loyalist Americans,* 34.

326 **They would offer** Bielinski, *An American Loyalist,* 25.

326 **"our log house"** Roger Morris to Mary Philipse Morris, June 11, 1776, *Letters,* 64.

327 **"You can not be insensible"** Judd, "Frederick Philipse III of Westchester County," in East and Judd, *Loyalist Americans,* 35.

328 **The British army apparently lived well** Callahan, *Royal Raiders,* 79.

328 **"handsome & fashionable"** Roger Morris to Mary Philipse Morris, August 16, 1777, *Letters,* 143.

328 **"total ignorance of the character of the refugees"** Wertenbaker, *Father Knickerbocker Rebels,* 214–15.

329 **"able and willing to bear arms"** Van Buskirk, *Generous Enemies,* 144.

329 **"Marther the Negro wench"** Miscellaneous receipts, LC microfilm reel 117, cited in Grieff, 123.

330 **"silver tissue"** Baumgarten, 31.

331 **"the most perfect hilarity"** Wertenbaker, 201–2.

332 **"We were near what was called Phillip's House"** Hufeland, *Westchester County,* 264–65.

333 **"intent to subvert the government"** The New York Act of Attainder, or Confiscation Act. 22d of October, 1779, *Greenleaf's Laws of New York from the First to the Twentieth Session Inclusive,* 1:26.

334 **"The woman is returned"** Beckwith to Marquard, Morris's House, July 28, 1781, Sir Henry Clinton Papers, 167:10, William L. Clements Library, University of Michigan.

336 **"worthy inhabitants of Philipsburg"** Van Buskirk, 166.

336 **Then each household** Grieff, 84–85.

Afterword

340 **"The Octagonal Room"** *New York Packet,* May 26, 1785, cited in Grieff, 142.

343 **"one of the finest Estates"** John Enys, *The American Journals of Lt. John Enys,* ed. by Elizabeth Cometti (New York: Adirondack Museum and Syracuse University Press, 1976), 192–93.

Sources

Books

Adams, Arthur G. *The Hudson Through the Years*. 1983. Reprint, New York: Fordham University Press, 1996.

———, ed. *The Hudson River in Literature: An Anthology*. New York: Fordham University Press, 1988.

Aguet, Isabelle. *A Pictorial History of the Slave Trade*. Geneva: Editions Minerva, 1971.

Allison, Charles Elmer. *The history of Yonkers: from the earliest times to the present. . . .* New York: Wilbur B. Ketcham, 1896.

Apthecker, Herbert et al. *American Negro Slave Revolts*. New York: International Publishers, 1993.

Armour, David Arthur. *The Merchants of Albany, New York, 1686–1760*. New York: Garland, 1986.

Axtell, James. *Beyond 1492: Encounters in Colonial North America*. New York: Oxford University Press, 1992.

———. *The Invasion Within: The Contest of Cultures in Colonial North America*. New York: Oxford University Press, 1985.

Bachman, Van Cleaf. *Peltries or Plantations: The Economic Policies of the Dutch West India Company in New Netherland, 1623–1639*. Baltimore: Johns Hopkins University Press, 1969.

Barlow, Elizabeth. *The Forests and Wetlands of New York City*. Boston: Little, Brown, 1971.

Barnes, Donna R., and Peter G. Rose. *Matters of Taste: Food and Drink in Seventeenth-Century Dutch Art and Life*. Albany, NY: Albany Institute of History & Art, Syracuse University Press, 2002.

Baumgarten, Linda. *What Clothes Reveal: The Language of Clothing in Colonial and Federal America*. New Haven, CT: Yale University Press, 2002.

Beach, S. A. *The Apples of New York: Report of the New York Agricultural Experiment Station for the Year 1903*. Albany, NY: J. B. Lyon, 1905.

Beckles, Hilary. *A History of Barbados: From Amerindian Settlement to Nation-State*. Cambridge: Cambridge University Press, 1999.

———. *Natural Rebels: A Social History of Enslaved Black Women in Barbados*. New Brunswick, NJ: Rutgers University Press, 1989.

Belden, Louise Conway. *The Festive Tradition: Table Decoration and Desserts in America, 1650–1900*. New York: W. W. Norton/Winterthur Books, 1983.

Benson, Adolph B., trans. and ed. *Peter Kalm's Travels in North America: The English Version of 1770*. New York: Wilson-Erickson, 1937.

Berkin, Carol. *First Generations: Women in Colonial America*. New York: Hill and Wang, 1996.

Bielinski, Stefan. *An American Loyalist: The Ordeal of Frederick Philipse III*. Division of Historical Services, New York State Museum, 1976.

Biemer, Linda B. "Criminal Law and Women in New Amsterdam and Early New York." In *A Beautiful and Fruitful Place: Selected Rensselaerswijck Seminar Papers*. ed. Nancy Anne McClure Zeller. Albany, NY: New Netherland Publishing, 1991.

Blackburn, Roderic H., and Ruth Piwonka. *Remembrance of Patria: Dutch Arts and Culture in Colonial America, 1609–1776*. Albany, NY: Albany Institute of History and Art, 1988.

Bliven, Bruce, Jr. *Under the Guns: New York: 1775–1776*. New York: Harper & Row, 1972.

Bolton, Reginald. *Indian Paths in the Great Metropolis*. New York: Museum of the American Indian, Heye Foundation, 1922.

Bone, Kevin, ed. *The New York Waterfront: Evolution and Building Culture of the Port and Harbor*. New York: The Monacelli Press, 1997.

Bonomi, Patricia U. *A Factious People: Politics and Society in Colonial New York*. New York: Columbia University Press, 1971.

Booth, Sally Smith. *Seeds of Anger: Revolts in America, 1607–1771*. New York: Hastings House, 1977.

Borland, Bruce, ed. *The Social Fabric: American Life from 1607–1877*. New York: Harper-Collins, 1995.

Boxer, Charles R. *The Dutch Seaborne Empire, 1600–1800*. New York: Knopf, 1965.

Boyle, Robert H. *The Hudson River: A Natural and Unnatural History.* New York: W. W. Norton, 1969.

Braun, Esther K., and David P. Braun. *The First Peoples of the Northeast.* Lincoln, MA: Lincoln Historical Society, 1994.

Bridenbaugh, Carl, ed. *Gentleman's Progress: The Itinerarium of Dr. Alexander Hamilton 1744.* Chapel Hill: University of North Carolina Press, 1948.

Brodhead, John Romeyn, Esq. *Documents Relative to the Colonial History of the State of New-York; Procured in Holland, England and France.* Albany, NY: Weed, Parsons, 1853.

Burger, Carl. *Beaver Skins and Mountain Men: The Importance of the Beaver in the Discovery, Exploration, and Settlement of the North American Continent.* New York: Dutton, 1968.

Burke, Thomas E., Jr. *Mohawk Frontier: The Dutch Community of Schenectedy, New York, 1661–1710.* Ithaca, NY: Cornell University Press, 1991.

Burnaby, Rev. Andrew. *Travels Through the Middle Settlements in North-America, in the years 1759 and 1760 with Observations upon the State of the Colonies.* Ithaca, NY: Cornell University Press, 1960.

Burrows, Edwin G., and Mike Wallace. *Gotham: A History of New York City to 1898.* New York/Oxford: Oxford University Press, 1999.

Bushman, Richard L. *The Refinement of America: Persons, Houses, Cities.* New York: Knopf, 1992.

Callahan, North. *Royal Raiders: The Tories of the American Revolution.* Indianapolis, IN: Bobbs-Merrill, 1963.

Cantwell, Anne-Marie, and Diana diZerega Wall. *Unearthing Gotham: The Archaeology of New York City.* New Haven, CT: Yale University Press, 2001.

———. *Touring Gotham's Archaeological Past: Eight Self-Guided Walking Tours through New York City.* New Haven, CT: Yale University Press, 2004.

Carmer, Carl. *The Hudson.* New York: Farrar & Rhinehart, 1939.

Carse, Robert. *Ports of Call.* New York: Charles Scribner's Sons, 1967.

Chidsey, Donald Barr. *The Loyalists: The Story of Those Americans Who Fought Against Independence.* New York: Crown, 1973.

Cohen, Paul E., and Robert T. Augustyn. *Manhattan in Maps 1527–1995.* New York: Rizzoli, 1997.

Colden, Cadwallader. *Papers relating to an act of the Assembly of the province of New-York, for encouragement of the Indian trade, etc. and for prohibiting selling of Indian goods to the French. . . .* New York: William Bradford, 1724.

Condon, Thomas J. *New York Beginnings: The Commercial Origins of New Netherland.* New York: New York University Press, 1968.

Cordingly, David. *Under the Black Flag: The Romance and the Reality of Life Among the Pirates.* New York: Harcourt Brace, 1995.

Couzens, M. K., compiler. *Index of Grantees of Lands Sold by the Commissioners of Forfeitures of the Southern District of the State of New York. . . .* M. K. Couzins, 1880.

Crary, Catherine S., ed. *The Price of Loyalty: Tory Writings from the Revolutionary Era.* New York: McGraw-Hill, 1973.

Sources

Cronon, William. *Changes in the Land: Indians, Colonists, and the Ecology of New England.* New York: Hill and Wang, 1983.

Crowley, John E. *The Invention of Comfort: Sensibilities and Design in Early Modern Britain and Early America.* Baltimore: Johns Hopkins University Press, 2001.

Cruikshank, Helen G., ed. *John and William Bartram's America: Selections from the Writings of the Philadelphia Naturalists.* New York: Devin-Adair, 1957.

Cunnington, Cecil Willett, and Phillis Cunnington. *The History of Underclothes.* Mineola, NY: Dover, 1992.

Cushman, Elisabeth. *Historic Westchester, 1633–1933.* Tarrytown, NY: Westchester County Publishers, 1933.

Dankers, Jasper, and Peter Sluyter. *Journal of a Voyage to New York and a Tour in Several of the American Colonies in 1679–80.* Translated by Henry C. Murphy. Brooklyn: Long Island Historical Society, 1867.

Davis, Thomas J. "These Enemies of Their Own Household: Slaves in 18th Century New York. " In *A Beautiful and Fruitful Place: Selected* Rensselaerswijck *Seminar Papers.* ed. Nancy Anne McClure Zeller. Albany, NY: New Netherland Publishing, 1991.

Davis, T. J. *A Rumor of Revolt: The "Great Negro Plot" in Colonial New York.* New York: Free Press, 1985.

De Bonneville, Francoise et al. *The Book of Fine Linen.* Paris: Flammarion, 1994.

De Pauw, Linda Grant, and Conover Hunt. *"Remember the Ladies": Women in America, 1750–1815.* New York: Studio Books/Viking Press, 1976.

Deetz, James. *In Small Things Forgotten: An Archaeology of Early American Life.* 1977. Reprint, New York: Anchor Books, 1996.

Delaney, Edmund T. *New York's Turtle Bay Old & New.* Barre, MA: Barre Publishers, 1965.

Denton, Daniel. *A Brief Description of New York: Formerly Called New Netherlands.* Cleveland: Burrows Brothers, 1902.

Dexter, Elisabeth Anthony, Ph.D. *Colonial Women of Affairs: A Study of Women in Business and the Professions in America Before 1776.* Boston: Houghton Mifflin, 1924.

Dilliard, Maud Esther. *An Album of New Netherland.* New York: Twayne, 1963.

Douglas, Emily Taft. *Remember the Ladies: The Story of Great Women Who Helped Shape America.* New York: G. P. Putnam's Sons, 1966.

Druett, Joan. *Hen Frigates: Wives of Merchant Captains Under Sail.* New York: Simon & Schuster, 1998.

———. *She Captains: Heroines and Hellions of the Sea.* New York: Simon & Schuster, 2000.

Duffy, John. *A History of Public Health in New York City, 1625–1866.* New York: Russell Sage Foundation, 1968.

Dunn, Richard S. *Sugar and Slaves: The Rise of the Planter Class in the English West Indies, 1624–1713.* Chapel Hill: University of North Carolina Press, 1972.

Dunshee, Kenneth Holcomb. *As You Pass By.* New York: Hastings House, 1952.

Earle, Alice Morse. *Colonial Days in Old New York.* New York: Charles Scribner's Sons, 1896.

———. *Costume of Colonial Times.* 1894. Reprint, New York: Empire State, 1924.

————. *Curious Punishments of Bygone Days*. New York: H. S. Stone, 1896.

————. *Home Life in Colonial Days*. 1898. Reprint, Middle Village, NY: Jonathan David, 1975.

————. *Sun Dials and Roses of Yesterday*. . . . New York: Macmillan, 1902.

East, Robert A., and Jacob Judd, eds. *The Loyalist Americans: A Focus on Greater New York*. Tarrytown, NY: Sleepy Hollow Restorations, 1975.

Eberlein, Harold Donaldson. *The Manors and Historic Homes of the Hudson Valley*. Philadelphia: J. B. Lippincott, 1924.

Fernow, B. *Documents relating to the history and settlements of the towns along the Hudson and Mohawk rivers*. . . . Albany, NY: Weed, Parsons, 1881.

Fish, Hamilton, LL.D. *New York State: The Battleground of the Revolutionary War*. New York. Vantage Press, 1976.

Fisher, Leonard Everett. *The Hatters*. New York: Franklin Watts, 1965.

Fisher, Sidney G. *Men, Women & Manners in Colonial Times*, vol. 2. 1897. Reprint, Detroit: Singing Tree Press, 1969.

Gehring, Charles, trans. and ed. *Fort Orange Records, 1656–1678*. Syracuse, NY: Syracuse University Press, 2000.

Gehring, Charles T., and Nancy Anne McClure Zeller, eds. *Education in New Netherland and the Middle Colonies: Papers of the 7th Rensselaerswyck Seminar of the New Netherland Project*. Albany, NY: New Netherland Project, New York State Library, 1985.

Gehring, Charles T., and William A. Starna, trans. and eds. *A Journey Into Mohawk and Oneida Country, 1634–1635: The Journal of Harmen Meyndertsz van den Bogaert*. Syracuse, NY: Syracuse University Press, 1988.

Glasse, Hannah et al. *The Art of Cookery Made Plain and Easy*. 1747. Reprint, Bedford, MA: Applewood Books, 1997.

Goodfriend, Joyce D. *Before the Melting Pot: Society and Culture in Colonial New York City, 1664–1730*. Princeton, NJ: Princeton University Press, 1992.

Gottesman, Rita S., ed. *The Arts and Crafts in New York, 1726–1776: Advertisements and News Items From New York City Newspapers*. New York: Da Capo Press, 1970.

Grant, Anne. *Memoirs of an American Lady, with sketches of manners and scenes in America as they existed previous to the Revolution*. 1808. Reprinted with unpublished letters and a memoir of Mrs. Grant by James Grant Wilson. New York: Dodd, Mead, 1901.

Grieff, Constance M. *The Morris-Jumel Mansion: A Documentary History*. Princeton, NJ: Heritage Studies, 1995.

Gunn, Fenja. *The Artificial Face: A History of Cosmetics*. Worcester, UK: David & Charles, 1973.

Haak, Bob. *The Golden Age: Dutch Painters of the Seventeenth Century*. New York: Harry N. Abrams, 1984.

Hale, Nathaniel Claiborne. *Pelts and Palisades; The Story of Fur and the Rivalry for Pelts in Early America*. Richmond, VA: Dietz Press, 1959.

Hall, Edward Hagaman. *Philipse Manor Hall*. New York: The American Scenic and Historic Preservation Society, 1930.

Hansen, Harry. *North of Manhattan: Persons and Places of Old Westchester.* New York: Hastings House, 1950.

Harrison, Molly. *The Kitchen in History.* New York: Charles Scribner's Sons, 1972.

Hauptman, Laurence M., and Jack Campisi, eds. *Neighbors and Intruders: An Ethnohistorical Exploration of the Indians on Hudson's River.* Ottawa: National Museums of Canada, 1978.

Heckewelder, Rev. John. *The History, Manners, and Customs of the Indian Nations Who Once Inhabited Pennsylvania and the Neighboring States.* Philadelphia: The Historical Society of Pennsylvania, 1876.

Henry, Alexander. *Travels and Adventures in Canada and the Indian Territories Between the Years 1760 and 1776.* Rutland, VT: Charles E. Tuttle, 1969.

Hodges, Graham Russell, and Alan Edward Brown, eds. *"Pretends to Be Free": Runaway Slave Advertisements from Colonial and Revolutionary New York and New Jersey.* New York and London: Garland, 1994.

Hodges, Graham Russell. *Root & Branch: African Americans in New York and East Jersey, 1613–1863.* Chapel Hill: University of North Carolina Press, 1999.

Holliday, Carl. *Woman's Life in Colonial Days.* 1922. Reprint, Williamstown, MA: Corner House, 1968.

Homberger, Eric, and Alice Hudson, cartographic consultant. *The Historical Atlas of New York City: A Visual Celebration of Nearly 400 Years of New York City's History.* New York: Henry Holt, 1994.

Hooker, Mark T. *The History of Holland.* Westport, CT: Greenwood Press, 1999.

Horsmanden, Daniel. *The New York Conspiracy.* 1744. Reprint, Boston: Beacon Press, 1971.

Hufeland, Otto. *Westchester County During the American Revolution, 1775–1783.* White Plains, NY: Westchester County Historical Society, 1926.

Huizinga, J. H. *Dutch Civilization in the Seventeenth Century and Other Essays.* New York: Frederick Ungar, 1968.

Innes, J. H. *New Amsterdam and Its People: Studies, Social and Topographical, of the Town Under Dutch and Early English Rule,* vols. 1, 2. 1902. Reprint, Port Washington, NY: Ira J. Friedman, 1969.

Innis, Harold A. *The Fur Trade in Canada: An Introduction to Canadian Economic History.* New Haven, CT: Yale University Press, 1930.

Irving, Washington. *Knickerbocker's History of New York,* ed. by Anne Carroll Moore. New York: Frederick Ungar, 1928.

———. *Stories of the Hudson.* Harrison, NY: Harbor Hill Books, 1984.

Jameson, J. Franklin, gen. ed. *Narratives of New Netherland, 1609–1664.* New York: Elibron Classics Replica Edition, Charles Scribner's Sons, 1909.

Janvier, Thomas. *In Old New York: A Classic History of New York City.* 1894. Reprint, New York: St. Martin's Press, 2000.

Jaray, Cornell, ed. *Historic Chronicles of New Amsterdam, Colonial New York, and Early Long Island.* Port Washington, NY: Ira J. Friedman, 1968.

Jenkins, Stephen. *The Greatest Street in the World: The Story of Broadway, Old and New, from the Bowling Green to Albany.* New York and London: G. P. Putnam's Sons, 1911.

Jennings, Francis. *The Founders of America: How Indians Discovered the Land, Pioneered in It, and Created Great Classical Civilizations, How They Were Plunged Into a Dark Age by Invasion and Conquest, and How They Are Reviving.* New York: W. W. Norton, 1993.

Johnson, Kathleen Egan. *The Limner's Trade: Selected Colonial and Federal Paintings from the Collection of Historic Hudson Valley.* Tarrytown, NY: Historic Hudson Valley, 1996.

Judd, Jacob, and Irwin H. Polishook, eds. *Aspects of Early New York Society and Politics.* Tarrytown, NY: Sleepy Hollow Restorations, 1974.

Kammen, Michael. *Colonial New York: A History.* New York: Oxford University Press, 1975.

Kauffman, Henry J. *The American Fireplace: Chimneys, Mantelpieces, Fireplaces & Accessories.* New York: Galahad Books, 1972.

Kennedy, Roger G. *Architecture, Men, Women and Money, 1600–1860.* New York: Random House, 1985.

Kenney, Alice P. *Stubborn for Liberty: The Dutch in New York.* Syracuse, NY: Syracuse University Press, 1975.

Kenney, Peter M., Frances Gruber Safford, and Gilbert T. Vincent. *American Kasten: The Dutch-Style Cupboards of New York and New Jersey, 1650–1800.* New York: Metropolitan Museum of Art, 1991.

Ketchum, Richard M. *Divided Loyalties: How the American Revolution Came to New York.* New York: John MacRae Books/Henry Holt, 2002.

Kilpatrick, William Heard. *The Dutch Schools of New Netherland and Colonial New York.* New York: Arno Press/New York Times, 1969.

Kim, Sung Bok. *Landlord and Tenant in Colonial New York: Manorial Society, 1664–1775.* Chapel Hill: University of North Carolina Press, 1978.

Knight, Sarah Kemble. *The Journal of Madam Knight.* Written in 1704. Reprint, Boston: David R. Godine, 1971.

Kraft, Herbert C., ed. *The Archaeology and Ethnohistory of the Lower Hudson Valley and Neighboring Regions: Essays in Honor of Louis A. Brennan.* Bethlehem, CT: Archaeological Services, 1991.

———. *The Lenape: Archaeology, History, and Ethnology.* Newark, NJ: New Jersey Historical Society, 1986.

Lepore, Jill. *New York Burning: Liberty, Slavery, and Conspiracy in Eighteenth-Century Manhattan.* New York: Knopf, 2005.

Linebaugh, Peter, and Marcus Rediker. *The Many-Headed Hydra: Sailors, Slaves, Commoners, and the Hidden History of the Revolutionary Atlantic.* Boston: Beacon Press, 2000.

Ludlum, David. *Early American Winters, 1604–1820.* Lancaster, PA: Lancaster Press, 1966.

Maddocks, Melvin. *The Atlantic Crossing.* Alexandria, VA: Time-Life Books, 1981.

Mark, Irving. *Agrarian Conflicts in Colonial New York, 1711–1775.* Port Washington, NY: Irving J. Friedman, 1965.

Martin, Calvin et al. *Keepers of the Game: Indian-Animal Relationship and the Fur Trade.* Los Angeles: University of California Press, 1978.

Matson, Cathy. *Merchants & Empire: Trading in Colonial New York.* Baltimore: Johns Hopkins University Press, 1998.

McManus, Edgar J. *A History of Negro Slavery in New York.* Syracuse, NY: Syracuse University Press, 1966.

Meeske, Harrison. *The Hudson Valley Dutch and Their Houses.* Fleischmanns, NY: Purple Mountain Press, 2001.

Merwick, Donna. *Possessing Albany, 1630–1710: The Dutch and English Experiences.* New York: Cambridge University Press, 1990.

Millar, John F. *American Ships of the Colonial & Revolutionary Periods.* New York: W. W. Norton, 1978.

Mintz, Sidney W. *Sweetness and Power: The Place of Sugar in Modern History.* New York: Viking/Elizabeth Sifton Books, 1985.

Mondfeld, Wolfram. *Historic Ship Models.* New York: Sterling, 1989.

Montgomery, Florence M. *Textiles in America, 1650–1870.* New York: W. W. Norton, 1984.

Moss, Roger W., ed. *Paint in America: The Colors of Historic Buildings.* New York: John Wiley & Sons/The Preservation Press, 1994.

Murphy, Henry C. *Anthology of New Netherland: Translations from the Early Dutch Poets of New York with Memoirs of Their Lives.* 1865. Reprinted, Port Washington, NY: Ira J. Friedman, 1969.

Nammack, Georgiana C. *Fraud, Politics, and the Dispossession of the Indians: The Iroquois Land Frontier in the Colonial Period.* Norman, OK: University of Oklahoma Press, 1969.

Norton, Mary Beth. *Liberty's Daughters: The Revolutionary Experience of American Women, 1750–1800.* Boston: Little, Brown, 1980.

Norton, Thomas Elliot. *The Fur Trade in Colonial New York, 1686–1776.* Madison: University of Wisconsin Press, 1974.

O'Callaghan, E. B. *The Documentary History of the State of New-York.* Albany, NY: Weed, Parsons, 1849–1851.

———. *History of New Netherland; or, New York under the Dutch.* 2 vols. New York: D. Appleton, 1846–1848.

Perl, Lila. *Slumps, Grunts, and Snickerdoodles: What Colonial America Ate and Why.* New York: Clarion Books/Seabury Press, 1975.

Phillips-Birt, Douglas. *Fore & Aft Sailing Craft: The Development of the Modern Yacht.* London: Seeley, Service, 1962.

Phipps, Frances. *Colonial Kitchens, Their Furnishings, and Their Gardens.* New York: Hawthorn Books, 1972.

Postma, Johannes Menne. *The Dutch in the Atlantic Slave Trade, 1600–1815.* Cambridge: Cambridge University Press, 1990.

Rawley, James A. *The Transatlantic Slave Trade.* New York: W. W. Norton, 1981.

Rebora, Carrie, and Paul Staiti, Erica E. Hirsler, Theodore E. Stebbins Jr., Carol Troyen. *John Singleton Copley in America*. New York: The Metropolitan Museum of Art, 1995.

Rink, Oliver A. *Holland on the Hudson: An Economic and Social History of Dutch New York*. Ithaca, NY: Cornell University Press, 1986.

Ritchie, Robert C. *The Duke's Province: A Study of New York Politics and Society, 1664–1691*. Chapel Hill: University of North Carolina Press, 1977.

Rogozinski, Jan. *Honor Among Thieves: Captain Kidd, Henry Every, and the Pirate Democracy in the Indian Ocean*. Mechanicsburg, PA: Stackpole Books, 2000.

Rose, Peter G., trans. and ed. *The Sensible Cook: Dutch Foodways in the Old and the New World*. Syracuse, NY: Syracuse University Press, 1989.

Rosebrock, Ellen Fletcher. *Counting-house Days in South Street: New York's Early Brick Seaport Buildings*. New York: South Street Seaport Museum, 1975.

Rowley, Anthony. *The Book of Kitchens*. Paris: Flammarion, 2000.

Salinger, Sharon V. *Taverns and Drinking in Early America*. Baltimore: Johns Hopkins University Press, 2002.

Sandoz, Mari. *The Beaver Men: Spearheads of Empire*. 1964. Reprint, Lincoln: University of Nebraska Press, 1978.

Schama, Simon. *The Embarrassment of Riches: An Interpretation of Dutch Culture in the Golden Age*. New York: Vintage Books, 1997.

Scharf, J. Thomas, ed. *History of Westchester County*. New York: L. E. Preston, 1886.

Schaw, Janet. *Journal of a Lady of Quality; Being the Narrative of a Journey from Scotland to the West Indies, North Carolina, and Portugal, in the years 1774 to 1776*. New Haven, CT: Yale University Press, 1934.

Schecter, Barnet. *The Battle for New York*. New York: Walker, 2002.

Scholten, Catherine M. *Childbearing in American Society, 1650–1850*. New York: New York University Press, 1985.

Shelton, William Henry. *The Jumel Mansion*. Boston and Cambridge, MA: Houghton Mifflin and Riverside Press, 1916.

Shonnard, Frederic, and W. W. Spooner. *History of Westchester County*. 1900. Reprint, New York: Harbor Hill Books, 1974.

Shorto, Russell. *The Island at the Center of the World: The Epic Story of Dutch Manhattan, the Forgotten Colony that Shaped America*. New York: Doubleday, 2004.

Singleton, Esther. *Dutch New York*. New York: Dodd, Mead, 1909.

———. *Social New York Under the Georges 1714–1776*, vols. 1, 2. Port Washington, NY: Ira J. Friedman, 1902, 1969.

Skinner, Alanson. *The Indians of Manhattan Island and Vicinity*. New York: American Museum of Natural History, 1932.

Snow, Dean R., Charles T. Gehring, and William A. Starna, eds. *In Mohawk Country: Early Narratives about a Native People*. Syracuse, NY: Syracuse University Press, 1996.

Spectorsky, A. C., ed. *The Book of the Sea*. New York: Grosset & Dunlap, 1954.

Squire, Geoffrey. *Dress Art and Society, 1560–1970*. London: Studio Vista, 1974.

Stilgoe, John. *Common Landscape of America, 1580 to 1845.* New Haven, CT: Yale University Press, 1982.

Stokes, I. N. Phelps. *The Iconography of Manhattan Island, 1498–1909.* New York: Arno Press, 1967.

Thomas, Hugh. *The Slave Trade: The Story of the Atlantic Slave Trade, 1440–1870.* New York: Simon & Schuster, 1997.

Tieck, William A. *Riverdale, Kingsbridge, Spuyten Duyvil: New York City, A Historical Epitome of the Northwest Bronx.* Old Tappan, NJ: F. H. Revell, 1968.

Tree, Ronald. *A History of Barbados.* 1972. Reprint, London: Granada, 1981.

Trelease, Allen W. *Indian Affairs in Colonial New York: The Seventeenth Century.* Lincoln: University of Nebraska Press, 1997.

Tunis, Edwin. *Colonial Living.* Baltimore: Johns Hopkins University Press, 1999.

Van Buskirk, Judith L. *Generous Enemies: Patriots and Loyalists in Revolutionary New York.* Philadelphia: University of Pennsylvania Press, 2002.

Van Kirk, Sylvia. *Many Tender Ties: Women in Fur-Trade Society, 1670–1870.* Norman, OK: University of Oklahoma Press, 1989.

Van der Donck, Adriaen, ed. With an introduction by Thomas F. O'Donnell. *A Description of the New Netherlands.* Syracuse, NY: Syracuse University Press, 1968.

Van der Zee, Henri and Barbara. *A Sweet and Alien Land: The Story of Dutch New York.* New York: Viking Press, 1978.

Van Deursen, A. T. *Plain Lives in a Golden Age: Popular Culture, Religion and Society in Seventeenth-Century Holland.* Cambridge: Cambridge University Press, 1991.

Van Rensselaer, Mrs. John King. *The Goode Vrouw of Mana-ha-ta at Home and in Society 1609–1760.* New York: Charles Scribner's Sons, 1898.

Van Rensselaer, Mrs. Schuyler. *History of the City of New York,* vols. 1, 2. New York: Macmillan, 1909.

Van Tienhoven, Cornelis. "Information Relative to the Taking Up Land in New Netherland, 1650." *Documentary History of the State of New-York,* vol. 4. Albany, NY: Weed, Parsons, 1849–1851.

Venema, Janny. *Beverwijck: A Dutch Village on the American Frontier, 1652–1664.* Albany, NY: State University of New York Press, 2003.

Vetare, Margaret. *Philipsburg Manor Upper Mills.* Tarrytown, NY: Historic Hudson Valley, 2004.

Walton, Frank L. *Pillars of Yonkers: The Story of a Community from Tomahawks to Television.* New York: Stratford House, 1951.

Warwick, Edward, Henry C. Pitz, and Alexander Wyckoff. *Early American Dress: The Colonial and Revolutionary Periods.* New York: Benjamin Blom, 1965.

Wertenbaker, Thomas Jefferson. *Father Knickerbocker Rebels: New York City During the Revolution.* New York: Cooper Square, 1969.

Wertz, Richard W., and Dorothy C. Wertz. *Lying-In: A History of Childbirth in America.* New York: Schocken Books, 1977.

Sources

Wilcoxen, Charlotte. *Seventeenth Century Albany: A Dutch Profile*. Albany, NY: Albany Institute of History and Art, 1984.

Williams-Myers, A. J. *Long Hammering: Essays on the Forging of an African American Presence in the Hudson River Valley to the Early Twentieth Century*. Trenton, NJ: African World Press, 1994.

Wolf, Stephanie Grauman. *As Various as Their Land*. Fayetteville, AR: University of Arkansas Press, 2000.

Wolfert's Roost: Portrait of a Village. Irvington-on-Hudson, NY: Washington Irving Press, 1971.

Wolley, Charles A. *Two-Years' Journal in New York and Part of Its Territories in America*. 1701. Reprint, with an introduction and notes by Edward Gaylord Bourne, Harrison, NY: Harbor Hill Books, 1973.

Wood, William. *New England's Prospect*. Amherst: University of Massachusetts Press, 1977.

Wraxall, Peter, ed. With an introduction by Charles Howard McIlwain. *An abridgment of the Indian affairs contained in four folio volumes, transacted in the colony of New York, from the year 1678 to the year 1751*. Cambridge: Harvard University Press, 1915.

Wright, Louis B. *The Cultural Life of the Colonies*. 1957. Reprint, Mineola, NY: Dover, 2002.

Writer's Program of the Work Projects Administration. *A Maritime History of New York*. New York: Doubleday, Doran, 1941.

Zacks, Richard. *The Pirate Hunter: The True Story of Captain Kidd*. New York: Hyperion, 2002.

Zeller, Nancy Anne McClure, ed. *A Beautiful and Fruitful Place: Selected Rensselaerswijk Seminar Papers*. Albany, NY: New Netherland Publishing, 1991.

Articles

Andrews, Wayne, ed. "A Glance at New York in 1697: The Travel Diary of Dr. Benjamin Bullivant." *The New-York Historical Society Quarterly* 60 (January 1956).

Bakker, Cees. "Along the Spice Trails: Dutch Overseas Expansion in the 17th and 18th Centuries." *De Halve Maen* 66, no. 1 (Spring 1993).

Cornell, Thomas C. "Some Reminiscences of the Old Philipse Manor House in Yonkers and Its Surroundings." *Yonkers Historical Bulletin* 34, no. 1. 1988–1989.

Denton, Kevin A. "The American Kas." *De Halve Maen*.

Frijhoff, Willem. "New Views on the Dutch Period of New York." *De Halve Maen*, no. 2 (Summer 1998).

Galloway, J. H. "The Role of the Dutch in the Early American Sugar Industry." *De Halve Maen* (Summer 2003).

Goodfriend, Joyce D. "The Souls of African American Children: New Amsterdam." *Common-Place* 3, no. 4 (July 2003).

Hayes, John T. "Liberty Belles: Home Town Girls." *The Westchester Historian* 75, no. 1 (Winter 1999).

Sources

Horne, Field, ed. "Voices from Eighteenth Century Westchester." *The Westchester Historian* 77, no. 4 (Fall 2001).

John, F. Wallace. "The Building of Manor Hall." *Yonkers Historical Bulletin* 10, no. 2 (December 1963).

Judd, Jacob. "Frederick Philipse and the Madagascar Trade." *The New-York Historical Society Quarterly* 55, no. 4 (October 1971).

Lamb, Martha J. "The Philipse Manor-House." *Appleton's Journal* 11, no. 262 (March 1874).

Maika, Dennis J. "The Credit System of the Manhattan Merchants in the Seventeenth Century, Part I." *De Halve Maen* 63, no. 2 (June 1990).

————. "The Credit System of the Manhattan Merchants in the Seventeenth Century, Part II." *De Halve Maen,* 63, no. 3 (September 1990).

————. "The Credit System of the Manhattan Merchants in the Seventeenth Century, Part III." *De Halve Maen* 64, no. 1 (Spring 1991).

————. "Slavery, Race, and Culture in Early New York." *De Halve Maen* 73, no. 2 (Summer 2000).

Marcus, Leonard. "What's in a Street Name?" *Seaport* 12, no. 3 (Fall 1978).

Narrett, David E. "From Mutual Will to Male Prerogative: The Dutch Family and Anglicization in Colonial New York." *De Halve Maen* 65, no. 1 (Spring 1992).

Pessa, Joanna. "Fact or Fiction? Philipse Manor Hall in Literature." *Friends of Philipse Manor Hall News* 3, no. 2 (January 2003).

Piwonka, Ruth. "Old Pewter/Bright Brass: A Suggested Explanation for Conservativism in Dutch Colonial Culture." *De Halve Maen* 68, no. 2 (Summer 1995).

Randolph, Howard S. "The Hardenbrook Family." *New York Genealogical Society* (April 1939).

Schriek, C. R."The Philipse Jewel: A Legend Is Born." *De Halve Maen* 67, no. 2 (Summer 1994).

Shaw, Susanah, "New Light from Old Sources: Finding Women in New Netherland's Courtrooms." *De Halve Maen* 74, no. 1 (Spring 2001).

"They, Too, Lived in Philipse Manor Hall: Reminiscences of Latter-Day People Who Once Called the Historic Mansion Home." *Yonkers Historical Bulletin* 6, no. 1 (March 1958).

Van Zwieten, Adriana E. "'[O]n her woman's troth': Tolerance, Custom, and the Women of New Netherland." *De Halve Maen* 72, no. 1 (Spring 1999).

Voorhees, David William. "'how ther poor wives do, and are delt with': Women in Leisler's Rebellion." *De Halve Maen* 70, no. 2 (Summer 1997).

Walton, Frank L. "The Wading Place." *Yonkers Historical Bulletin* (July 1968).

Other Sources

Agnew, Aileen Button. "Silent Partners: The Economic Life of Women on the Frontier of Colonial New York." PhD diss., University of New Hampshire, 1998.

————. *Women and Property in Colonial New York: The Transition from Dutch to English Law, 1643–1727.* Ann Arbor, MI: UMI Research Press, 1983.

Brawer, Catherine Coleman, gen. ed. *Many Trails: Indians of the Lower Hudson Valley.* Catalog of exhibit at The Katonah Gallery, March 13–May 22, 1983.

Elliott, Carol. *Story of Philipsburg.* Working paper, Sleepy Hollow Restorations, Tarrytown, NY, 1966.

Garrison, J. Ritchie. "Philipse Manor Hall: An Historical Architectural Context and Stylistic Analysis." Prepared for the New York State Office of Parks, Recreation and Historic Preservation, Taconic Region, 2005.

Grumet, Robert Steven. "'We Are Not So Great Fools': Changes in Upper Delawaran Socio-Political Life, 1630–1758." PhD diss., Rutgers University, 1979.

Haley, Jacquetta M. "Philipsburg Manor Tricentennial 1693–1993." Research report, Historic Hudson Valley, Tarrytown, NY, June 1992.

————. "The Slaves of Philipsburg Manor, Upper Mills." Research report, Historic Hudson Valley, Tarrytown, NY, September 26, 1988.

Judd, Jacob, ed. "18th Century Life in America." Research report, Historic Hudson Valley, Tarrytown, NY, n.d.

————. "The State of Religion in Colonial New York." Research report, Historic Hudson Valley, Tarrytown, NY, n.d.

————. "Thanksgiving in New York." Research report, Historic Hudson Valley, Tarrytown, NY, 1958–59.

Historical Archeology at Philipse Manor Hall State Historic Site. State of New York leaflet, undated.

Huey, Paul R. "Archaeological Testing at Philipse Manor Hall, Yonkers, N.Y., 1994." New York State Office of Parks, Recreation and Historic Preservation, Research Unit, Bureau of Historic Sites, Peebles Island, Waterford, NY, March 1996.

Maika, Dennis J. "Commerce and Community: Manhattan Merchants in the Seventeenth Century." PhD diss., New York University, 1995.

————. "Philipse Family Commerce, 1650–1750." Research report, Historic Hudson Valley, Tarrytown, NY, October 1996 (Revised November 1997).

Morris, Roger, to Mary Philipse Morris. Letters 1755–1777. Morris-Jumel Mansion, New York, New York.

Otto, Paul Andrew. "New Netherland Frontier: Europeans and Native Americans Along the Lower Hudson River, 1524–1664." PhD diss., Indiana University, 1995.

Research Staff, Sleepy Hollow Restoration. "Chronological Report of the Activities of Frederick Philipse I, Frederick Philipse II, and Adolph Philipse." Tarrytown, NY, n.d.

Schaefer, Richard G. "Life in the 17th-Century Netherlands." *World Archaeological Congress 4,* University of Cape Town (January 10–14, 1999).

Schneir, Miriam. "'She Merchants' in 17th Century Holland and New Netherland." Paper presented at a seminar of the Program in Sex Roles and Social Change, Center for the Social Sciences, Columbia University, NY, April 1980.

Shattuck, Martha Dickinson, Ph.D. "A Civil Society: Court and Community in Bever-wijck, New Netherland, 1652–1664." PhD diss., Boston University, 1993.

Wendell, Evert. Account books, 1695–1758. The New-York Historical Society.

Wentworth, Ann Kennedy. "Woman of Business or Lady of the Manor?: An Archaeo-logical Examination of Changes in Gender Roles Among the Hudson Valley Elite During the Eighteenth Century." Paper, South Carolina Institute of Archaeology and Anthropology, University of South Carolina, 1995.

Index

Index

Index

Index

Index